Jennifer Russell

FOREST INSECTS
Principles and Practice
of Population Management

POPULATION ECOLOGY: Theory and Application
Series Editor: **Alan A. Berryman**
Washington State University
Pullman, Washington

POPULATION SYSTEMS: A General Introduction
Alan A. Berryman

FOREST INSECTS: Principles and Practice of Population Management
Alan A. Berryman

FOREST INSECTS
Principles and Practice of Population Management

Alan A. Berryman

Washington State University
Pullman, Washington

Plenum Press • New York and London

Library of Congress Cataloging in Publication Data

Berryman, A. A. (Alan Andrew), 1937–
  Forest insects.

  (Population ecology)
  Bibliography: p.
  Includes index.
  1. Forest insects. 2. Forest insects — Control. 3. Forest insects — Ecology. 4. Insect
populations. I. Title. II. Series.
SB761.B47   1986   595.7'052642   86-4890
ISBN 0-306-42196-8

First Printing — June 1986
Second Printing — November 1989

© 1986 Plenum Press, New York
A Division of Plenum Publishing Corporation
233 Spring Street, New York, N.Y. 10013

Printed in the United States of America

DEDICATION

As a boy in England, I dreamed of becoming a forester but my attempts to enter the competitive British forestry establishment were unsuccessful. As the next best alternative, my mentor at Cornwall Technical College, Gordon Ince, recommended entomology, so I went to London to become an entomologist. After graduating from the Royal College of Science, and at the urging of Professor O. W. Richards, I obtained a research assistantship from Professor Robert Usinger to work on the behavior of bed bugs at the University of California, Berkeley. It was during my first years at the University of California that I came into contact with that special breed, the forest entomologist, and more importantly with their leader at Berkeley, Professor Ronald Stark.

This book is dedicated to the fortuitous circumstances that led me to practice my chosen vocation (entomology) in the environment I always loved (forests), and to the people who guided me along this path. It is also dedicated to all my forest entomologist friends and colleagues who provided the stimulation, challenge, and comradeship that have made this journey so enjoyable.

PREFACE

This book is intended as a general text for undergraduates studying the management of forest insect pests. It is divided into four parts: insects, ecology, management, and practice.

Part I, Insects, contains two chapters. The first is intended to provide an overview of the general attributes of insects. Recognizing that it is impossible to adequately treat such a diverse and complex group of organisms in such a short space, I have attempted to highlight those insectan characteristics that make them difficult animals to combat. I have also tried to expose the insects' weak points, those attributes that make them vulnerable to manipulation by human actions. Even so, this first chapter will seem inadequate and sketchy to many of my colleagues. Ideally, this book should be used in conjunction with a laboratory manual covering insect anatomy, physiology, biology, behavior, and classification in much greater depth—in fact, this is how I organize my forest entomology course. It is hoped that this first chapter will provide nonentomologists with a general feel for the insects and with a broad understanding of their strengths and weaknesses, while Chapter 2 will provide a brief overview of the diverse insect fauna that attacks the various parts of forest trees and their products.

Part II, Ecology, deals with the interactions and interdependencies among forests, herbivorous insects, and insectivores, with the objective of understanding why pest outbreaks arise. The intimate relationship between insects and trees is explored in Chapter 3, and an attempt is made to explain how these interactions affect the dynamic structure of forest communities. Then, in Chapter 4, the question of why normally green forests suddenly turn red or gray as the result of insect depredations is addressed by studying the principles of population dynamics, a discipline that attempts to explain the causes of population changes. In the last chapter of Part II, the interactions between herbivorous insects and their predators, parasites, and pathogens are explored, with particular emphasis on the roles of these insectivores in regulating the abundance of pest insects.

Part III deals with the strategies and tactics of forest insect management. In Chapter 5 methods are described for monitoring ongoing insect infestations and forecasting future trends. In Chapter 6, the problem of assessing the risk of insect outbreaks by observing site and stand conditions is addressed with particular attention to the development of risk models. Then, in Chapter 8, the strategy of outbreak prevention is considered with reference to silvicultural and natural enemy manipulations that help maintain pest populations at suboutbreak levels. In Chapter 9 the basic principles and tactics of outbreak containment and suppression are described and, finally, an approach to pest management decision-making is presented in Chapter 10.

The last part of this book, Part IV (Chapter 11), is devoted to the practice of forest insect management. Drawing on analogies with medicine, the practice of forest insect management is defined as the art of diagnosis, prognosis, and prescription. This practical approach is illustrated with a specific example, the European spruce bark beetle, and then two other examples are set up for the student to analyze. A list of important forest insects with some relevant references is also supplied for students who wish to gain further practical experience by analyzing specific insect problems.

The slant in this book is ecological and, in particular, toward the general theory and principles of population ecology. Some may consider this approach too demanding and esoteric for the undergraduate student. I disagree. It is my firm conviction that foresters without a general understanding of how insects interact with forests, and what causes their populations to explode to outbreak levels, are ill prepared to face their insect competitors. There are so many *potential* pests inhabiting the forests of the world that a totally empirical approach is impractical, not to mention intellectually sterile. The general approach advocated in this book should enable the forestry graduate to *think about,* and to *attack,* any insect problem anywhere in the world. For this reason I have avoided details of the biology and control of specific forest insects that encumber many forest entomology texts. These details are already covered in numerous excellent publications that address specific regional or national forest insect problems.

ACKNOWLEDGMENTS

This book could not have been written without the untiring patience and guidance of my mentors, the continual enthusiasm and encouragement of my students, and the critical evaluation of my peers. The book was read in its entirety by Professor R. W. Stark, Professor J. P. Vité, and Professor A. Bakke. Parts of it were read by R. D. Akre, R. A. Beaver, J. H. Borden, J. J. Brown, E. P. Catts, P.-J. Charles, E. Christiansen, G. T. Ferrell, G. E. Long, R. R. Mason, F. D. Morgan, R. H. Miller, J. A. Millstein, P. W. Miles, D. Mlinšek, K. F. Raffa, F. D. Rhoades, F. M. Stephen, and B. E. Wickman. In addition, the book was read by undergraduate forestry students taking my course Forest Entomology in 1982, 1983, and 1985. I am most grateful for the constructive criticism that has guided the evolution of this book.

I also received many suggestions for insects to be included in the book and in the list at the end of Chapter 11. In particular, contributions were made by R. A. Beaver, B. Bejer, J. H. Borden, C. Géri, J.-C. Grégoire, E. Haukioja, W. G. H. Ives, F. Kobayashi, S. Larsson, J. H. Madden, W. J. Mattson, P. Niemela, C. P. Ohmart, A. Roques, H. Schmutzenhofer, J. Selander, R. F. Shepherd, D. Wainhouse, W. G. Wellington, and R. H. Werner.

Many of the ideas presented in this book were forged from discussion and debate with my friends and colleagues. Many have contributed in this way, but I would particularly like to acknowledge, in addition to those mentioned above, W. Baltensweiler, R. N. Coulson, D. L. Dahlsten, A. P. Gutierrez, F. P. Hain, A. S. Isaev, R. W. Reid, L. Safranyik, J. A. Schenk, N. C. Stenseth, M. W. Stock, K. J. Stoszek, and H. S. Whitney. I hope those who I have missed will excuse my forgetfulness.

I would like to thank the following for permission to use figures and quotes: the United States Forest Service (Figs. 4.8, 6.1, 6.2, 6.4, 6.6, 6.9, 7.1, 10.3,

10.4, and the frontispiece illustrations for Parts II, III, and VI), the Society of American Foresters (Figs. 4.10, 6.10a, and 7.4), the Entomological Society of Canada (Figs. 5.8 and 5.12), Plenum Publishing Corporation (Fig. 3.5), Idaho Department of Lands (frontispiece for Part I), Reinhold Publishing Corporation (quote fronting Part I), Annual Reviews Incorporated (quote fronting Part II), the Pan-Pacific Entomologist (quote fronting Part III), and the Southern Baptists Convention's Radio and Television Commission (quote by Chief Seattle heading the epilogue). Finally, I thank G. H. Baker, M. H. Brookes, G. E. Daterman, C. J. DeMars, D. G. Fellin, M. M. Furniss, Y. J. Hardy, R. L. Livingston, J. H. Madden, R. R. Mason, A. Roques, A. R. Stage, and B. E. Wickman for providing drawings and prints, even though I did not use all of them, their efforts are appreciated.

Finally, I would like to thank Doris Lohrey-Birch and Jane Bower for their patience and perseverance in typing this manuscript, and the College of Agriculture and Home Economics at Washington State University for providing me with their assistance.

CONTENTS

PART II
ECOLOGY

PART III
MANAGEMENT

PART IV
PRACTICE

PART I

INSECTS

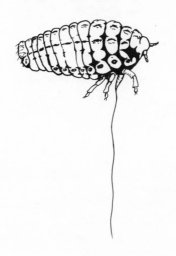

"to discover and exploit weak points in the structure, physiology, behavior and ecology of insects"
K. Graham, 1963, *Concepts of Forest Entomology,* Reinhold.

Drawing of the balsam woolly adelgid, *Adelges piceae,* by N. E. Johnson and the gouting it causes to fir twigs by R. L. Livingston.

CHAPTER 1

AN INTRODUCTION TO THE INSECTS

Insects are one of the most diverse groups of organisms ever to have crawled and flown over the face of the earth. Arising in the late Devonian era, more than 350 million years ago, insects have diversified and spread into every conceivable ecological niche; today there are over 1 million different species classified in more than two dozen separate orders (Fig. 1.1). Insects have successfully overcome the challenges of radical geologic and climatic changes including the successive ice ages. They will undoubtedly survive attacks by *Homo sapiens,* with whom they compete for food and fiber, as insects are a genetically flexible and adaptable group of animals. They will learn to live in our polluted wastelands and will survive our campaigns against them and against our fellows, for insects can exist in radiation levels 100 times greater than those that kill humans. How then can we compete with these masters of adaptability? Obviously we cannot eliminate them from the globe. But perhaps we can reduce their ravages to some tolerable level at which we can coexist in relative harmony.

Whenever we contemplate a foe, the first thing we should try to do is to understand him. How does he work? What are his strengths and weaknesses? Knowing these, we can attack his weak points and be cognizant of his ability to counterattack. The history of human struggle against insects is strewn with underestimates of their adaptive abilities. Perhaps the best known example is the "permanent solution" to the insect problem discovered in 1939—the insecticide DDT. During the years after its discovery, this potent insecticide was used to control numerous insect pests. The warning of the *thinking* biologist, that widespread and continuous use of this pesticide would result in selection of genotypes resistant to the chemical, was ignored by the pest control specialists. Soon many of our major insect pests had evolved resistance, and the insects had won another round. So let us try to understand the insects,—their strengths as well as their weaknesses.

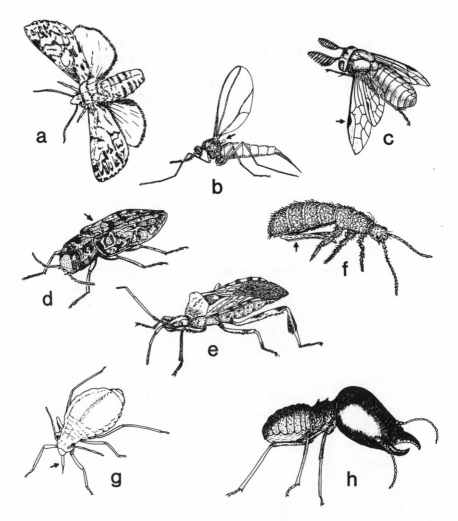

Figure 1.1. A few of the more common insect orders found in forests (not drawn to scale). (a) Pine moth, Order Lepidoptera. Note scales covering body and wings. (b) Cone gall midge, Order Diptera. Note hindwings reduced to small balancing organs or halters. (c) Sawfly, Order Hymenoptera. Note membranous wings with thickened vein or stigma on leading edge. (d) Metallic wood borer, Order Coleoptera. Note hardened protective forewings or elyra. (e) Leaf-footed bug, Order Hemiptera. Note forewings (hemielytra) half leathery, half membranous. (f) Springtail, Order Collembola. Note springing organ (furcula) and absence of wings. (g) Aphid, Order Homoptera. Note sucking mouthparts. (h) Soldier termite, Order Isoptera. Note wingless and blind. (a, b, d, e, g redrawn from R. W. Stark *et al.*, 1963.)

1.1. EXTERNAL STRUCTURE AND FUNCTION

Unlike us, with our internal supporting structure, insects have an external skeleton, or *exoskeleton*. Covered from head to foot by these hard plates, or *sclerites*, insects are much like medieval knights in armor. In order to move, the sclerites are joined by flexible membranes, giving them the jointed appendages that place them, with the lobsters, spiders, centipedes, and the like, into the phylum *Arthropoda* (meaning jointed legs) (Fig. 1.1).

The exoskeleton of an insect, also called the *integument*, or cuticle, is a laminated structure that is hard, yet flexible, and water impermeable. The outermost layer, the *epicuticle*, or wax layer, is made up of esters of long-chain fatty acids and alcohols and is extremely important in preventing water loss from the body cavity. Because insects are so small, they have a large surface area relative to their total volume, making evaporation and desiccation a serious problem. If the wax layer is rubbed off by abrasive materials, insects may desiccate and die. For example, abrasive volcanic ash descending after the explosion of Mount St. Helens in 1980 caused extensive mortality to many insect populations in eastern Washington.

The next layer, the *exocuticle*, is composed of the nitrogenous polysaccharide *chitin*, a very tough but flexible material, and a hardening protein called *sclerotin*. The process of hardening, or tanning, is called *sclerotization*. Beneath the exocuticle is an unsclerotized chitinous layer that forms the flexible *endocuticle*. Areas of the insect integument at the membranous joints are composed largely of endocuticle, whereas the hardened sclerites contain both the exo- and endocuticular layers. Beneath this last nonliving layer lies a row of living secretory cells, the *epidermis*, set on an extremely thin connecting tissue, the *basement membrane*. Most of the epidermal cells are secretory and give rise to the entire cuticle.

The integuments of insects bear numerous hairs and spines, some being articulated in cuticular sockets and some having tactile or chemosensory functions (i.e., they enable the insect to feel, smell, and taste the surrounding environment). Although the insect exoskeleton provides it with protection from many dangerous chemicals, physical damage, desiccation, pathogenic invasions, and so forth, it also poses some problems. In particular, body growth is only possible through shedding the integument (*molting*) and secreting a new one, with growth occurring before the hardening process takes place. During molting, the insect is vulnerable to physical, chemical, and biological elements.

The rigid integument of insects forms an ideal surface for structural modification and coloration; many insects have structural forms and colors resembling thorns, twigs, leaves, and bark flakes—they are experts at camouflage. Others have developed bright, flashy coloration which warn predators of their nasty taste, or are flashed suddenly to frighten attackers away. Still others have evolved forms which mimic, with great detail, their dangerous or distasteful relatives. The advantage of a fly mimicking the form, and even sound, of a wasp or bee is immediately apparent. All in all, the insect integument forms a very effective means of support and protection. Perhaps the only disadvantage is its weight, which may be one of the principal reasons why insects have not evolved to the size of vertebrates.

Insects arose more than 350 million years ago from wormlike ancestors which looked more like soft-bodied centipedes than our present-day insects. The ancestral body form was tubelike and made up of a number of segments, each bearing a pair of walking legs. During the course of evolution, the form of the body has changed considerably, and all but three pairs of the walking legs have been lost or have been modified for purposes other than walking. The body of the modern insect is divided into three parts: the *head, thorax,* and *abdomen* (Fig. 1.1). The head forms a discrete articulated unit, the head capsule, which is made from the fusion of several ancestral body segments. The appendages of these segments have been lost or modified into structures that surround the mouth and are used for manipulating (*maxillae* and *labium*) and masticating (*maxillae* and *mandibles*) the insect's food (see the large mandibles of the soldier termite in Fig. 1.1h). However, these components of the primitive insect chewing mouth parts have been modified and changed in some groups into lapping (bees), sponging (certain flies), siphoning (butterflies and moths), and piercing–sucking organs (certain flies and true bugs) (Fig. 1.1g). The head also bears the important sensory structures, such as a pair of large *compound eyes* through which the insect perceives color and form. The compound eyes are made up of a large number of distinct and independent hexagonal units, or *ommatidia,* and on the front and top of the head are additional simple light sensors (the *ocelli*), which help the insect detect differences in light intensity. Also protruding forward from the head are a pair of sensory organs (*antennae*), which serve as tactile sensors and detectors of chemical stimuli in the environment, particularly odors emanating from their food, or from other members of the species (e.g., sex attractants, trail chemicals). Antennae come in many shapes and sizes and are often used for classifying insects (see the complex antennae of the male sawfly in Fig. 1.1c).

The insect head is attached to a three-segmented thorax that bears all the locomotory appendages, three pairs of walking legs (two per segment) and, usually, two pairs of wings (on the second and third segments) (Fig. 1.1). The reduction of the walking appendages from the ancestral multiple-legged form to six articulated appendages is a remarkable example of adaptive efficiency. A six-legged stance provides maximum walking stability with the least number of appendages, each step forming a tripod composed of the front and hind legs on one side together with the middle leg on the other. Triangular support structures are well known for their efficiency, stability, and strength. Together with claws, suction pads, and the like on the terminal tip of the leg (*tarsus*), the insects are able to attach themselves to, and balance on, almost any surface. Insect legs have also adapted to perform many functions, from the grasshopper's enlarged jumping leg, and the praying mantid's grasping foreleg, to the remarkable hind leg of the honey bee with pollen basket, rakes, and brushes.

Long before birds and bats evolved, the carboniferous jungles hummed with the flight of giant dragonflies and other primitive insects. The ability to fly, perhaps more than any other single attribute, was probably responsible for the incredible adaptive success of the class Insecta. Flight provides the insects with many advantages, from the location of rare and temporary habitats, to escaping unfavorable environments and the persecution of their enemies. Insect swarms arriving unexpectedly in field and forest, where they were previously absent or in low numbers, has plagued and frustrated human crop and animal husbandry from time immemorial, such as the devastating migrations of desert locusts. Certainly the flight ability of insects is one of the major characteristics that makes them so difficult to combat.

The mechanics of insect flight is itself a remarkable process. Unlike birds and bats, whose wings are modified walking appendages, insect wings are extensions of their exoskeleton. Wing movement is brought about, indirectly, by distortion of the thoracic integument caused by powerful longtitudinal and vertical musculature. The rapid contractions of these muscles, together with the rigid springiness of the thoracic integument, enables most insects to achieve wing-beat frequencies around 30–40/sec. Some of the more highly evolved groups, especially flies, have muscles that contract many times for each nerve impulse allowing them to attain wing-beat frequencies of up to 1000/sec. Hence the incredible maneuverability and darting speed of the hover fly.

Insect wings are formed of membranous and flexible cuticle supported by a system of veins (Fig. 1.1c). Like the other external structures, there are many modifications of insect wings and venation, some of which form useful characters for taxonomic identifications. For example, the wings of moths and butterflies (Lepidoptera) are covered with minute overlapping, and often brightly colored scales (Fig. 1.1a); the hind wings of flies (Diptera) are reduced to tiny balancing organs called *halteres* (Fig. 1.1b); the forewings of beetles (Coleoptera) form a hard sclerotized protective covering (*elytron*) for the delicately folded hindwings (Fig. 1.1d); and some insects do not have wings (Fig. 1.1f); or the wings are absent in some forms (Fig. 1.1g,h).

The third major division of the insect body, the abdomen, is made up of 11 segments, some of which may be invaginated or lost in the process of adaptation, and contains most of the internal organs and reproductive structures. The external reproductive organs are often complex, acting as a sort of lock-and-key arrangement that, being peculiar to the species, prevents interspecific crossbreeding. For this reason, the external genitalia are often of value for taxonomic identification. In some insects, such as the mayflies, the tip of the abdomen bears three long filamentous extensions composed of two *cerci* and a central caudal filament.

1.2. INTERNAL STRUCTURE AND FUNCTION

The internal structures of insects are associated with digestion of food, excretion of wastes, circulation of nutrients and hormones, movement of oxygen, transfer of information, reproduction, and musculature. The *digestive system* of the insects forms a convoluted tube starting with the mouth, where *salivary glands* aid in predigestion, and ending with the anus, where waste material is ejected as feces. Although there are many modifications, the fore part of the gut is primarily concerned with mastication and storage, the mid-part with digestion, and the hind-part with excretion. In many species the mid-gut contains microorganisms that aid in the digestion of certain materials, such as the digestion of cellulose by protozoa in termite guts. At the junction of the *midgut* and *hindgut* are a series of tubes, the *Malpighian tubules,* which function to remove waste products from the blood (analogous to our kidneys). the terminal *rectum* has an important function in removing water from the feces before expulsion. Most insects also excrete uric acid rather than urea, another adaptation that conserves water (uric acid is insoluble).

The *circulatory* fluid (*hemolymph*) of insects, unlike that of vertebrates, is not enclosed in tubes but rather fills the body cavity (*hemocoel*) and bathes all the internal organs. A single pulsating organ, the *dorsal vessel* or "heart," moves fluids forward and, with the help of dorsal and ventral diaphragms, ensures a slow circulation. The blood of insects does not function primarily as a vehicle for oxygen transport but rather for the movement of nutrients and hormones. Among the other organs in the insect hemocoel is the so-called *fat body,* which stores and metabolizes energy reserves (analogous to the verterbrate liver).

Oxygen is carried to the internal tissues through a network of cuticular tubes called *trachaeae* and their associated air sacs. Starting as openings (*spiracles*), a pair of which are usually present on each thoracic and abdominal segment, the trachaeae ramify throughout the body ending in minute *tracheoles.* Movement of oxygen is largely by diffusion, aided to some degree by muscular contractions and body movement. The limitations of this system, in terms of moving oxygen over long distances, is one of the constraints on the size of insects.

The *nervous system* of insects consists of a paired ventral *nerve cord* linked to interconnected segmental *ganglia,* from which nerve fibers extend to the various organs. In the head region the nerve chord circles the digestive tract and enlarges dorsally to form the *brain.* Associated with the nervous system, particularly the brain, are various glands that secrete hormones into the hemocoel.

The *endocrine system* of insects produces hormones that control the processes of growth, metamorphosis, and reproduction. For example, insect molting is initiated by a group of hormones called *ecdysones.* These are secreted by thoracic glands after they have been stimulated by brain hormones. Ecdysones also induce metamorphosis of the immature forms into the adult when other hormones, called *juvenile hormones,* are in low concentration. Juvenile hormones, which are released from brain glands (*corpora allata*), inhibit the development of adult characteristics and, therefore, maintain the insect in the juvenile stages. Another type of juvenile hormone, produced in the adult insect, stimulates egg development. Hormones are also involved in diapause initiation, sclerotization of the exoskeleton, pheromone production, and probably other physiological processes.

The internal reproductive organs of insects consist of the male *testes* and female *ovaries* together with their accessory tubes and glands. Although reproduction is usually bisexual, *parthenogenesis* (unisexual) reproduction is not uncommon, particularly among aphids (Homoptera) and some parasitic wasps (Hymenoptera). Parthenogenetic reproduction is a mechanism which permits single females to colonize a favorable environment very quickly and naturally creates severe problems for the pest manager.

Insects are usually very fecund organisms, some females laying up to 1000 eggs in their short lifetimes. This tremendous reproductive potential also creates difficulties, for the growing population can exploit available food resources in a very short timespan. In addition, because genetic recombinations and reassortment occur during sexual reproduction, enormous numbers of genetically diverse individuals are produced, giving insect populations great *genetic diversity* with which to adapt to environmental changes, including those instituted by human actions. Hence many insects have evolved *resistance* to both synthetic and natural chemical pesticides.

The *musculature* of insects is attached to protrusions (*apodemes* and *apophyses*) on the inside of the integument. Because of the external, tubelike support, the muscles can obtain exceptional leverage so that, size for size, insects are among the strongest animals on earth. An ant, for example, can carry a leaf or another insect several times its body weight—in its jaws!

1.3. BIOLOGY AND BEHAVIOR

Because insects, like all other invertebrates, are cold blooded (*poikilothermic*), and because it is necessary for them to shed their exoskeleton in order to grow, insect life cycles are characterized by discrete changes in form (*metamorphosis*) and are strongly influenced by seasonal temperatures. In our discussion of insect structure and function we concentrated on the adult because it is the most complex and variable form of development. However, in order to reach the adult stage, insects have to pass through a series of developmental stages, starting with the *egg*. In almost all insects, the eggs are deposited by the female on or in leaves or twigs, the ground, or other animals. In a few species, such as the tsetse fly, however, the young develop within the mother's body. Insect eggs are of many shapes and sizes, some are smooth and round, others ribbed and sculptured. They may be laid singly or in groups. After a variable period of time, depending on external conditions of temperature and moisture, the egg hatches into the first motile stage.

Insects have two major modes of *postembryonic development*. In the more primitive groups, the newly hatched individual has a form and structure very similar to the adult, except that the wings and sexual organs are absent or undeveloped (Fig. 1.2a). As the young insect or *nymph* grows, it passes through a series of molts, molting being governed by hormonal secretions from the endocrine system. As it grows, wing pads and sexual organs develop and it gradually assumes the adult form. This kind of development is known as *simple metamorphosis* because no radical changes in form take place during development. On the other hand, the more advanced orders go through a series of discrete and specialized structural changes known as *complex metamorphosis* (Fig. 1.2b).

In these groups a highly specialized feeding stage, the *larva,* hatches from the egg and proceeds to grow through a series of *instars,* each instar being seperated by a molt. Although the larva may take various shapes and sizes it is an extremely simple, almost embryonic creature whose main function is to feed and store energy for the final transition to the adult. Not surprisingly, it is insect larvae that cause most of the damage to food and fiber, feeding on or within leaves, needles, twigs, trunks, roots, cones, seeds, and timber products. In order to transform from an uncomplicated "feeding bag" into a highly complex adult, the insect has to pass through a *pupal stage,* in which the extraordinary processes of tissue reorganization and reformation occur (Fig. 1.2b). The insect pupa has limited mobility and is, therefore, extremely vulnerable to predators, parasites, and other environmental hazards, including man. For this reason, pupae are often

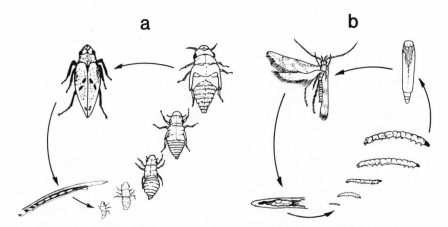

Figure 1.2. Insect growth and metamorphosis. (a) Simple metamorphosis of a spittle bug, *Aphrophora permutata* Uhler (Homoptera). (b) Complex metamorphosis of a leaf-mining caterpillar, *Coleotechnites starki* (Freeman) (Lepidoptera). (Redrawn after R. W. Stark *et al.,* 1963).

protected by a *cocoon* spun by the last larval instar (Lepidoptera and most Hymenoptera), or in some cases by the hardened skin of that last instar (Diptera). Many insects also pupate in the ground, in crevices and holes on trees, or in other protected locations sought out by the prepupal larval stage.

Emergence of the adult insect from the pupa is also a highly vulnerable time for those groups with complex metamorphosis. Before the newly emerged adult can fly, the wings have to be inflated by body pulsations which force fluid into the veins. After this, the adult is able to initiate its search for food and mates, and a place to lay its eggs.

The *adult* insect is an extremely efficient machine for sensing its environment and for responding to external stimuli though flight, mating, feeding and oviposition behaviors. Although the behavior of insects may be extremely complex, it is governed almost completely by innate genetic properties, or instinct (some, such as ants and bees do seem to have limited learning abilities). Thus, although the insects have remarkable behavioral adaptations, they act in a stereotyped, mechanical, and predictable manner. This characteristic is probably their greatest weakness and is being exploited more and more in the continuing battle against insect pests.

Although some insects reproduce *parthenogenetically* (asexually) at times, and males have never been observed in a few species, sexual reproduction is the norm in most species. Sexual reproduction ensures the genetic interchange that gives the insects their great *genetic plasticity*. Thus, although individual insect behavior is stereotyped and rigid, insects have tremendous abilities to adapt through natural selection over a number of generations.

Insects have evolved complex behavioral patterns associated with mate location and mating; these include *visual display, stridulation* (sound communication), and *sex pheromones* (chemical communication). Pheromones, in particular, have received considerable attention by scientists because of their potential for managing insect pests. Pheromones are volatile chemicals, frequently derived from compounds present in the insect's food, which are secreted into the environment for the purpose of communicating with other members of the same species; e.g., attracting mates (many insect groups), laying trails (ants), controlling development (caste determination in social insects), marking parasitized hosts (many parasitic wasps), guiding others to a resisting host in order to overwhelm its defenses (bark beetle aggregation pheromones), and repelling others from a food supply which is optimally colonized (bark beetle antiaggregation pheromones). The identification and synthesis of these chemicals is providing a useful tool in forest insect survey (Chapter 6) and offers potential in insect pest control (Chapter 9).

Because the larval stages of insects are flightless and have limited sensory capabilities, the adult females possess highly developed behavioral mechanisms for locating suitable food supplies for its offspring. Chemical cues, in the form of volatile attractants emanating from suitable hosts, and repellents from unsuitable ones, has been the main adaptive approach to achieve these ends. Once again, information on the host chemicals involved in this process can be extremely useful in the fight against insect pests; e.g., repellent chemicals can be sprayed on crops or introduced into the plant by breeding programs or, for that matter, attractive chemicals can be bred out.

Some female insects also require nourishment from their environment in order for their eggs to mature. Rich sources of carbohydrate or protein, such as *nectar* and *honeydew* (secretions from aphids and scale insects) are often used for this purpose. Chemical receptors on the antennae, or other body parts (e.g., the tarsi of house flies), enable them to respond to these stimuli.

Behavioral complexity has reached its xenith in the *social insects* (wasps, bees, ants, and termites). Provisioning and care of the young is the central theme in all these insect societies, and division of labor (social castes) controlled by pheromones is common in the more highly developed groups. Complex behaviors have also evolved for communicating the location of food sources (the dance of the honey bee), gathering food (the march of the army ant), and for building their intricate and environmentally controlled hives and nests (honey bees). Even altruistic behavior, that most exalted of social attributes, is evidenced in the self-sacrifice of the soldier ant in defense of the colony.

The highly efficient locomotory and sensory adaptations of the adult insect determine where the rather helpless larva will be born and therefore the suitability of its food supply and its exposure to parasites, predators, and disease. True, the larva does have certain behavioral adaptations governing feeding behavior, wind-borne dispersal, locating pupation sites, spinning cocoons, and defense against predators (e.g., some sawflies store repellent plant chemicals and secrete them when attacked), but their neurological capacity is rather limited when compared with the adult. Because of the critical importance of adult behavior, the life cycle is timed so that adults are present when these functions can best be performed. For example, adult emergence is timed for spring or early summer in species which feed on mature foliage (e.g., the European pine sawfly), where adults often fly and deposit diapause (overwintering) eggs in late summer or fall in those species whose larvae depend on buds or new foliage (e.g., Douglas-fir tussock moth). In other species, the flight period is synchronized to times when their hosts are most vulnerable to attack; e.g., the mountain pine beetle flies in midsummer when water stress on its host is most severe.

If adult behavior controls the most favorable place for its offspring to be born, it is processes occurring within the immature insect which determine the timing of adult eclosion. The timing of the insect life cycle is usually linked to unfavorable periods of the year. Insects are cold-blooded, and those living in extreme climates have to deal with a long period when temperatures are decidedly unfavorable. Hibernation is a common phenomenon in temperate animals, including insects, but it is a somewhat hazardous practice for cold-blooded organisms. This is because a warm spell in winter may allow the animal to resume activity at a time when there is no food available, and it may be followed, quite suddenly, by extreme cold. To circumvent this problem, many insects enter a state of suppressed metabolic activity, or *diapause,* which is only broken by warm weather if the insect has experienced a certain number of cold days. Some species, or genetic strains within a species, may remain in diapause for more than one winter; this adaptation is beneficial for those species that utilize resources that vary greatly from year to year (e.g., cone crops in forests).

The initiation of diapause is often dependent on day length rather than temperature. The length of the day, of course, is a more predictable sign of the onset of winter than is temperature, and this mechanism enables the insect to avoid the surprise "cold snap." The ability of insects to synchronize their physiology and behavior accurately in response to light is attributable to the presence of complex internal mechanisms, commonly called *biological clocks.* Biological clocks not only control seasonal rhythms, *via* diapause and other processes, but also control daily (*circadian*) rhythms that influence molting as well as various other activities such as feeding, flight, and mating.

Insects may enter diapause in any stage of development; e.g., the egg (gypsy moth), young larva (spruce budworm), mature larva (European pine shoot moth), pupa (fall webworm), and adult (Douglas-fir beetle). Hence, the *timing* of the insect life cycle is controlled by the stage that enters diapause and the length of the diapause period, while the *place* in which development takes place is controlled by the sensory and locomotory adaptations of the adult. This dual mechanism controlling the time and place of insect activity forms a basis for the design of life history strategies in the Insecta. Hence, insects with obligatory diapause will usually have a single generation a year, a common phenomenon in temperate regions. In more tropical environments we find species with several to many generations a year that hibernate only when conditions get too dry. Some temperate species have both obligatory and faculative diapausing strains, enabling them to reproduce an extra partial generation in favorable years. Still others have extended diapause, which spreads the risk over several years. Extended diapause is particularly prevalent among cone and seed insects, which have to survive several years when host trees are without new cones. The most highly adapted insects produce nondiapausing generations or diapausing generations depending on the day length to which the mother is exposed (many aphids). This final sophistication often leads to highly complex life cycles in which succeeding generations are very different in modes of reproduction and habits, often utilizing different host plants (Fig. 1.3).

Because many insects are genetically preconditioned to enter diapause, it may be considered a weak point in their life cycle. A method has been proposed to exploit this weakness by lengthening the daylight period with artificial lighting and so prevent the insect from entering diapause. These insects would then encounter winter cold in a vulnerable physiological state; i.e., changes in the

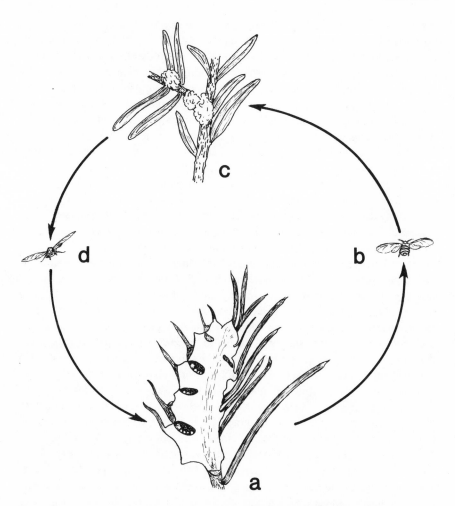

Figure 1.3. The complex 2-year life cycle and alternation of generations of a spruce aphid of the genus *Adelges* (Homoptera): (a) galls are formed on the primary host (spruce) in spring (here the gall cut open to show aphids in chambers); (b) winged females fly to secondary host (fir, Douglas-fir or larch) in late summer; (c) winged and wingless asexual generations are produced on secondary host (Aphid females covered by white, cottony tufts); (d) winged females fly to spruce in mid-summer where sexual generations are produced.

insect body during the diapausing process, such as the accumulation of antifreeze chemicals (glycerol and sorbitol) and the removal of water, enable the insect to survive temperatures as low as −50°C. Research on this potential method of insect control has shown that a strategically timed flash of light from a strobo-scope can achieve the desired effect of extending the daylength period.

Although the overall synchronization of the insect life cycle is determined to a large degree by the overwintering diapause strategy, the rate of development during the growing season is largely controlled by external factors. Developmental rates in poikilothermic insects are strongly influenced by temperature, but moisture also has an important modifying effect. In species with a rigid (obligatory) diapause, only one generation can be produced per year. However, weather conditions during the growing season will determine whether the insect is in the correct stage to enter diapause. Population collapses of some forest insects have been attributed to cooler than normal summer temperatures that delayed development and caused the insect to enter winter in a nondiapausing stage (e.g., the insects may have to spend winter in the larval stage when it is the adult that normally diapauses).

In species with facultative diapause, or which are nondiapausing, the number of generations produced per year is strongly related to weather conditions. Because the number of generations produced in a period of time has a tremendous influence on the rate of population growth, insect population explosions are often associated with warm, dry weather. Weather, of course, also affects other insect activities such as mating, oviposition, and flight. Flight, is particularly sensitive to weather because insect flight normally requires muscle temperatures in excess of 30°C, and rainfall poses a severe hazard for flying insects. Patterns of wind and weather, particularly storm fronts, may also have a dramatic influence on insect flight patterns and dispersal.

1.4. SUMMARY

We have seen that insects are an extremely well adapted and successful group of animals that form a formidable foe in the struggle for the earth's resources. Their many strengths give them advantages in this struggle including, but not limited to, the following:

1. Flight, which enables them to colonize widely dispersed food supplies, to escape the attacks of their enemies, and to find mates when their populations are sparse
2. High fecundity and fast rates of development, which enable them to rapidly exploit available food reserves and to maintain high genetic diversity
3. Genetic plasticity, through which their populations can rapidly adapt to changes in their environments, including those caused by human actions
4. Small size, enabling them to occupy diverse and protected habitats.

On the other side, insects have certain weaknesses that humans can exploit to their advantage, including the following:

1. Stereotyped behavior that causes them to respond automatically and irrevokably to certain stimuli, such as pheromones, light (diapause), and host odors

2. Susceptibility to abrasion of the integument and desiccation
3. Temperature/moisture-dependent development, which makes them vulnerable to environmental manipulation
4. Metamorphosis, which makes them vulnerable during transition stages and to hormones that control growth and molting.

If it seems that insect strengths outweigh their weaknesses, it should not be too surprising. The insects have survived and prospered for more than 300 million years. By comparison, the mammals have only been around for 200 million years or so and *Homo sapiens* for less than one million. Humans, however, have two extraordinary advantages in their struggle with the insects—intrinsic intelligence and enormous technological capabilities. If *Homo sapiens* has a serious weakness, it is a tendency to rely on technology rather than intelligence in the conflict with insects. Technology, however, can only supply tactical weapons. Strategy, the intelligent art of deciding when, where, and with what weapon to hit the enemy to achieve optimum effectiveness, requires a comprehensive understanding of the enemy. This understanding can only be obtained through basic knowledge of the biology, physiology, behavior, and ecology of insects. It is this accumulation of knowledge, and its intelligent use, that is our greatest weapon against the insects.

REFERENCES AND SELECTED READINGS

Atkins, M. D., 1978, *Insects in Perspective,* MacMillian, New York.

Beck, S. D., 1968, *Insect Photoperiodism,* Academic Press, New York.

Beroza, M., 1970. *Chemicals Controlling Insect Behavior,* Academic Press, New York.

Brown, F. A., 1972, The "clocks" timing biological rhythms, *Am. Sci.* **60**:756–766.

Edney, E. B., 1957, *The Water Relations of Terrestrial Arthropods,* Cambridge University Press, Cambridge.

Elizinga, R. J., 1978, *Fundamentals of Entomology,* Prentice-Hall, Englewood Cliffs, New Jersey.

Farb, P., and the editors of *Life,* 1962, *The Insects,* Time Inc., New York.

Frost, S. W., 1959, *Insect Life and Insect Natural History,* 2nd ed., Dover, New York.

Johnson, C. G., 1963, The aerial migration of insects, *Sci. Am.* **209**:132–138.

Salt, R. W., 1961, Principles of insect cold-hardiness, *Ann. Rev. Entomol.* **6**:55–74.

Stark, R. W., Graham, K., and Wood, D. L., 1973, *Manual of Forest Insects and Damage,* University of California, Berkeley.

Wells, M., 1968, *Lower Animals,* McGraw-Hill, New York.

Wilson, E. O., 1971, *Insect Societies,* Belknap Press of Harvard University, Cambridge, Massachusetts.

CHAPTER 2

FOREST INSECT PESTS

The term pest describes those organisms that have a negative impact on human survival or well-being, either acting as parasites; transmitting pathogens; competing with humans for food, fiber, or other useful resources; or just plain annoying them. The term pest, therefore, is highly subjective and reflects the human viewpoint. A truly objective and impartial view (i.e., a strictly scientific one) may lead to the conclusion that the concept of a pest is inappropriate, for all organisms play important roles in the development and maintenance of ecological communities (see Chapter 3). From this viewpoint we may even come to the conclusion that only one organism really threatens the stability and persistence of living systems on this planet—that *Homo sapiens* is the real pest!

On the other hand, we must realize that each species, as well as each individual of a species, is involved in a struggle with others for a limited supply of essential resources. From the subjective perspective of the species or individual, therefore, the idea of a competitor, or a pest, is a meaningful one. This book examines insects inhabiting forest environments from both points of view, for in order to fully understand insects and their role in the forest, we must adopt a scientific approach, while to understand human reaction to them we must view things more subjectively.

If we were to catalog all the insect species inhabiting a forested area, we would be surprised to find how many there were, as well as how few were actually considered a problem. We might then ask ourselves: What makes these few species so special that we consider them pests? In answering this question, we would first observe that many species do not, in fact, feed on resources that are valuable to man. Many are predaceous or parasitic on other insects, while others feed on decaying vegetation. These species may be thought of more correctly as allies because they aid in suppressing potential pests and in recycling nutrients back to the trees. Thus, the first and most obvious criterion for a forest pest is that it feeds on a commodity that man is interested in utilizing himself and thereby act as a direct competitor.

Even after narrowing it down to those species that feed on valuable trees, however, we would find few to be serious problems. This is because most of these species do not become abundant enough to cause serious economic damage. In other words, their numbers are so low that, from the manager's viewpoint, they cause negligible loss of wood fiber and other valuable resources (e.g., recreation, wildlife). Thus, an important characteristic of most insect pests is that they attain numbers, even if only occasionally or rarely, that cause serious economic losses. The economic aspect is important in this definition because greater losses can be tolerated in a low value crop (e.g., an inaccessible forest) than in one that is valuable (e.g., a plantation or seed orchard). Because numerical abundance is a critical attribute of most forest pests, the study of numerical change in insect populations, and an understanding of the causes underlying these changes, is essential for the forest pest manager (see Chapter 4).

Although most serious forest pest problems are the result of violent and often unpredictable population changes, we may also find a few species that are pests at even low populations levels. As mentioned earlier, a sparse insect population may be considered a problem if it affects extremely valuable resources, such as ornamental and shade trees or plantations and seed orchards, where considerable time and money have often been invested. However, there is one other type of insect which may become a problem even at low population densities, those that transmit pathogenic microorganisms or inject toxic materials into their hosts. For example, the smaller European elm bark beetle carries the fungus that causes Dutch elm disease, and the balsam woolly adelgid injects salivary toxins into fir trees. We note that both insects are not native to North America but were accidentally introduced from Europe. In fact, we will find that many forestry problems that are caused by pathogens (or toxins) transmitted by insects result from the introduction of the pathogen, the insect, or both from a foreign country. In Chapter 3 we will see that this problem results because tree and pathogen have not had the evolutionary time to adapt to each other.

Insects may cause economic damage to forest trees and their products in many ways, from death of the tree to growth reduction, reduced grade of timber, stem deformity, reduced seed crops, or may have more subtle effects on recreation, wildlife, esthetics, and fire hazard. In the remainder of this chapter we will briefly examine the major forest pest problems according to the part of the tree on which the insects feed.

2.1. INSECTS FEEDING ON CONES AND SEEDS

A number of different groups of insects attack the cones and/or seeds of forest trees. For example, small wasps penetrate the cone and seed coat with their long ovipositors and their larvae feed within the seed (Fig. 2.1a); adult cone beetles bore galleries in green cones causing them to abort before the seeds have developed (Fig. 2.1b); the larvae of cone moths bore through the cone and may destroy most of the seeds (Fig. 2.1c); leaf-footed bugs penetrate the cone with their piercing mouthparts to suck the juices from individual seeds (see Fig. 1.1e); flies lay eggs on the conelets and their maggots bore through the cone (Fig. 2.1d) or form galls on cone scales causing the seeds to abort (adult shown in Fig. 1.1b); cone weevil larvae bore through scales and seeds (Fig. 2.1e); and many other species feed incidentally on flowers, cones, and seeds.

Damage caused by cone and seed insects is proportional to the size of the insect population relative to the number of susceptible cones. In years when cone crops are small and populations of cone and seed insects are high, the entire crop may be destroyed. The abundance of cone and seed insects in any year seems to be most strongly affected by the size of the cone crop in previous years. This is not surprising because an abundance of food in one year should increase insect reproduction leading to a large population of offspring to attack cones the next year.

Cone production by most forest tree species growing under natural conditions is rather unpredictable. However, a bumper crop is often preceded by a series of poor cone production years and is usually followed by a very poor cone crop. Cone and seed insect populations and their damage, therefore, are usually associated with these dynamic aspects of cone production. For example, during years of low cone production, insect populations are limited by the amount of food present, but they increase considerably in years of high production. Even so, these bumper crops are so large that only a small percentage of the seed is lost. In other words, damage by seed and cone insects is usually inconsequential in years of heavy cone crops. In the following years, however, the entire crop may be lost as the enlarged insect population is forced to feed on a much reduced food supply.

The dynamics of cone production in commercial seed-producing areas (e.g., seed orchards) may be considerably altered by intensive management practices, such as tree spacing, pruning, and fertilization. Efforts to maintain high cone production in these areas may well result in larger insect populations and heavier damage. In addition, because of the expense incurred in managing seed production areas, even a 5 or 10% loss may be economically intolerable. Most efforts to control seed and cone insects are therefore confined to areas being intensively managed for seed production.

Problems from cone and seed insects may also arise when natural seeding is required following logging operations, e.g., seed tree cuts, where only a few select trees are left to provide seed. This problem is particularly acute if the cone crop is small and the insect population large. Conditions such as these may delay reforestation for several years, or necessitate planting operations, both of which increase the forest manager's costs.

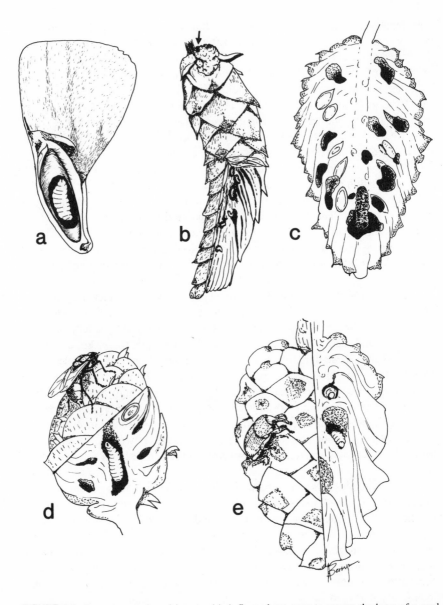

FIGURE 2.1. Some cone and seed insects. (a) A fir seed cut open to expose the larva of a seed chalcid (Hymenoptera, Torymidae: *Megastigmus* sp.). (b) A white pine cone opened to expose the larvae of a cone beetle (Coleoptera, Scolytidae: *Conophthorus* sp.). Note attack site of adult and boring dust at cone attachment. (c) A ponderosa pine cone cut open to show cavities excavated by a coneworm (Lepidoptera, Pyralidae: *Dioryctria* sp.). (d) A fly (Diptera, Anthomyiidae: *Hylemya* sp.) laying eggs under the scale of a larch cone; the sectioned cone is shown exposing the fly maggot and its feeding galleries. (e) A cone weevil (Coleoptera, Curculionidae: *Pissodes* sp.) preparing an oviposition site on a Scots pine cone; sectioned cone shows feeding grubs.

2.2. INSECTS FEEDING ON BUDS, SHOOTS, AND TWIGS

Insects that feed on or in buds or growing terminals are known to the forester as tip and shoot insects. Buds and growing terminals are attacked by many of the same insects that feed on cones; for example, cone beetles, coneworms, and weevils also feed on terminals. In addition, buds and shoots are attacked by bud moths, aphids, gall midges and twig-boring beetles. By far the most serious problems, however, are attributable to terminal weevils and shoot moths. The larvae of terminal weevils, such as the notorious white pine weevil (*Pissodes strobi*), mine within the shoots causing the growing terminals to die (Fig. 2.2a). Similarly, larvae of shoot moths often kill buds and leaders (Fig.

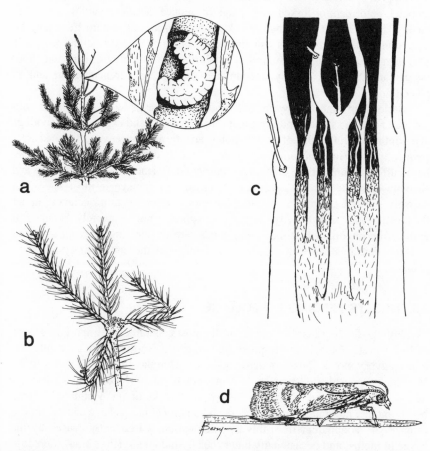

FIGURE 2.2. Some tip and shoot borers. (a) A lodgepole pine terminal killed by a terminal weevil (Coleoptera, Curculionidae: *Pissodes* sp.). See also Fig. 2.1e. (b) Distorted growth of a young pine after death of a terminal shoot due to shoot moth attack. (c) Deformed stems resulting from shoot moth infestation. (d) Adult pine shoot moth (Lepidoptera, Olethreutidae: *Rhyacionia* sp.).

2.2b). These insects usually attack the apical terminal of the tree causing irregular stem growth or multiple branching when secondary terminals take over dominance (Fig. 2.2b,c). Thus, trees which have been subjected to repeated attack by tip and shoot insects may have a stunted, bushy appearance, or at least, malformed or forked boles, and their value for timber production can be greatly reduced or eliminated. On the other hand, a certain amount of shoot damage may enhance the value of Christmas trees, which are sometimes manually pruned to attain the same dense, bushy appearance.

Damage from tip and shoot insects is usually most severe in plantations and in natural regeneration after heavy cutting. Attack on the buds and terminals of very young seedlings will often kill them, or so stunt their growth that they are overtopped by undesirable plant species. Attacks on saplings often result in crooked, deformed, or forked stems, and this reduces aesthetic values and timber quality (Fig. 2.2c,d). Although attack on pole sized and mature trees may be quite common, the impact is usually inconsequential because height growth is well advanced and bole form already established. Problems from tip and shoot insects are therefore usually encountered during forest regeneration and in plantations.

Serious infestations by tip and shoot insects seem to be associated with particular site and stand conditions that affect the vigor and exposure of seedlings and saplings. Outbreaks of shoot moths and terminal weevils, for instance, frequently occur in stands growing on dry or poorly drained sites, or on sites that have suffered compaction, erosion, or nutrient deficiencies. In addition, tip and shoot insects often prefer open conditions so that damage is most intense in sparsely stocked clearcut areas and plantations. Reproduction occurring under the forest canopy or in heavily stocked plantations, is not so severely damaged. It is therefore important that the manager use care in selecting planting sites and stocking densities, as well as avoiding damage to the site during harvesting operations.

2.3. INSECTS FEEDING ON FOLIAGE

Foliage-feeding insects are one of the most economically important groups of forest pests. Outbreaks of spruce budworms in North American coniferous forests, gypsy moths in eastern hardwoods, and Douglas-fir tussock moths in the West have received widespread public attention because of the extensive destruction caused to mature forests. In Europe, outbreaks of larch budmoths, pine beauty moths, nun moths, pine loopers and autumnal moths have caused similar damage. As is apparent from this list, defoliation is frequently caused by the larvae of moths, and occasionally butterflies (Lepidoptera) (Fig. 2.3a). Lepidopteran larvae may feed by browsing on the foliage (Fig. 2.3b), or mining inside the leaves or needles (Fig. 2.3c), and some may produce tent-like shelters and nests (Fig. 2.3d). Another important group, however, is the sawflies (Hymenop-

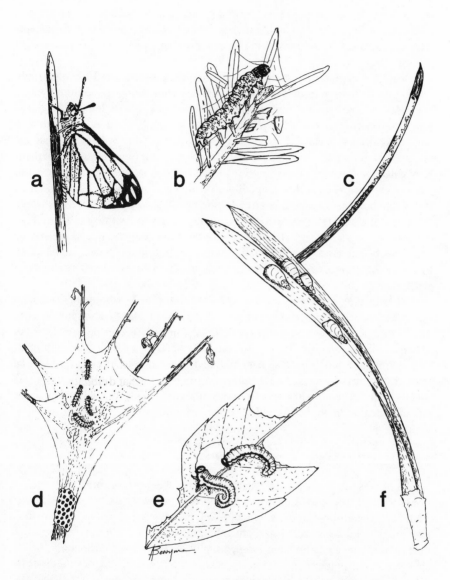

FIGURE 2.3. Some insects that feed on foliage. (a) Pine butterfly (Lepidoptera, Pieridae: *Neophasia menapia*) laying eggs on a pine needle. (b) western budworm caterpillar (Lepidoptera: Tortricidae: *Choristoneura* sp.). (c) Lodgepole needleminer (Lepidoptera, Gelechiidae: *Coleotechnites* sp.). See also Fig. 1.2b. (d) The nest of a tent caterpillar (Lepidoptera, Lasiocampidae: *Malacosoma* sp.). Note egg mass at the base of the nest. (e) Larvae of the striped alder sawfly (Hymenoptera, Tenthredinidae: *Hemichroa* sp.). (f) The pine needle scale (Homoptera, Diaspididae: *Chionaspis* sp.).

tera) whose larvae often feed in colonial groups, on the foliage of conifers and hardwoods (Fig. 2.3e). In addition, leaf beetles (Coleoptera), leafcutting ants (Hymenoptera), and scale insects (Homoptera) (Fig. 2.3j) defoliate forest trees on occasion.

With the exception of the sucking insects (Homoptera and Hemiptera), the feeding of defoliating insects removes the photosynthesizing tissue from the tree and reduces carbohydrate production. The immediate effect of defoliation, therefore, is a reduction in the vigor and growth of the tree. Reduction in growth may have a significant economic impact on timber production when large areas are affected. Defoliation sometimes results in considerable tree mortality, particularly when the forest has been subjected to other stress factors such as nutrient or water deficiencies, extreme competition, or old age. In addition, the weakened trees often become susceptible to tree-killing insects, such as bark beetles, which frequently cause extensive mortality following defoliator outbreaks. Of course, the survival of forest trees also depends on their innate properties as well as on the intensity and duration of defoliation. For example, some species can be killed by a single complete defoliation (e.g., western hemlock), while others can withstand several years of defoliation (e.g., most hardwoods).

Leaf-feeding insects attack all ages of forest trees but outbreaks are often associated with older stands, overstocked stands, or stands growing on poor sites. For example, spruce budworm outbreaks in eastern Canada usually occur in dense, mature stands of balsam fir, while Douglas-fir tussock moth outbreaks in the West and pine processionary outbreaks in southern France, seem to be more severe on drier sites. Many entomologists believe that defoliator outbreaks usually occur in stands suffering from physiological stress. We will examine the rationale behind this thinking more thoroughly in Chapters 3 and 7.

2.4. INSECT FEEDINGS ON TRUNKS

A number of insect groups feed on or within the stems of forest trees but, of these, the bark beetles (Coleoptera, Scolytidae) are by far the most destructive. Bark beetle adults bore into the bark of living or recently killed trees and lay their eggs in galleries constructed in the cambium region (Fig. 2.4a,b). At the same time they introduce pathogenic fungi, usually of the genus *Ceratocystis,* which spread through phloem and xylem, blocking the tree's transport systems (Fig. 2.4c). The newly attacking adults also produce powerful pheromones that, together with resins exuding from the tree, attract others of the same species. It is this combined attack by many beetles, together with the spread of pathogenic fungi that causes the death of all, or part (e.g., top or strip kills) of the tree. Because pheromones also concentrate the flying beetle population in an area, dead trees often occur in groups. In some species, such as the southern pine beetle, "group kills" may enlarge to cover several hectares.

Outbreaks of bark beetles occur at irregular intervals but are usually associ-

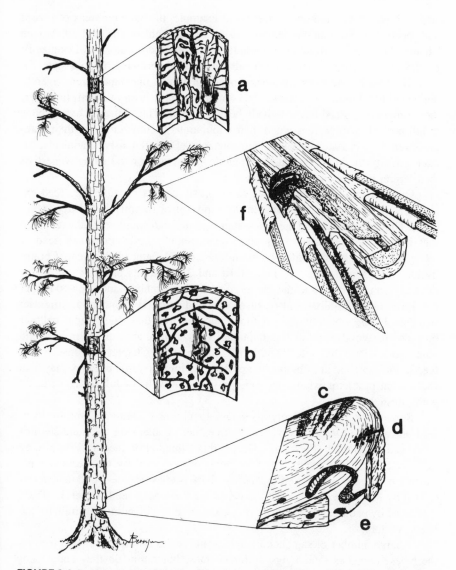

FIGURE 2.4. Some insects attacking the stems of pines. (a) Pine engravers (Coleoptera, Scolytidae: *Ips pini*) bore galleries in the phloem of small trees or the tops of large pines, particularly logging or thinning residues. (b) Southern pine beetles (Coleoptera, Scolytidae: *Dendroctonus frontalis*) prefer to bore their winding galleries in the mid- and lower bole of living pines. (c) The sapwood of trees killed by bark beetles is usually stained blue by fungi of the genus *Ceratocystis*. (d) Adult ambrosia beetles (Coleoptera, Scolytidae: *Gnathotricus* sp.) bore galleries in the sapwood of recently dead trees; their larvae feed on fungi growing in the side chambers, or "cradles." (e) Roundheaded wood borers (Coleoptera, Cerambycidae) tunnel in the phloem and xylem of weak trees and logs. (f) Adult pine shoot beetles (Coleoptera, Scolytidae: *Tomicus* sp.) feed in the terminals of healthy pines before breeding in logs or very weak trees.

ated with stands that are under stress. For example, the huge outbreak of spruce bark beetles in Scandinavia started during the drought of the mid-1970s (see Chapter 11), while outbreaks of mountain pine, southern pine, and Douglas-fir beetles often start in overstocked stands, or in stands growing on poor sites. Fir engraver beetle outbreaks frequently occur in stands that have been severely defoliated by leaf-eating insects. Once outbreaks have been initiated, however, they sometimes spread into relatively healthy stands. This ability of bark beetles to kill normal healthy trees when their populations become large, or their "aggressiveness", is associated with their tolerance of host defensive chemicals, the pathogenicity of their associated symbiotic fungi, and/or the attractiveness of their aggregation pheromones (Table 2.1).

Other scolytid beetles, known as ambrosia beetles because they feed exclusively on fungi, bore galleries in the sapwood of recently killed trees (Fig. 2.4d). They bore holes and the black-staining fungi can cause serious degrade to logs in the woods or awaiting processing in lumber yards. Logs and dead or dying trees are also degraded by roundheaded and flatheaded borers (Cerambycidae and Buprestidae; see Figs. 1.1d and 3.3). Adults of these insects lay their eggs on the bark and the larvae mine in the cambium and sapwood (Fig. 2.4e). Degrade of lumber is also caused by woodwasps (Hymenoptera, Siricidae; see Fig. 11.9), which introduce fungi into the tree that cause rapid cellular breakdown. Woodwasps in Australia and New Zealand are a problem in exotic pine stands where they attack and kill living trees (see Chapter 11). Insects that feed in the sapwood are obviously a problem in lumber yards but they may also cause legal problems when they emerge from the walls of houses constructed with infested lumber.

Some bark beetles (Scolytidae) and longhorned beetles (Cerambycidae) feed on twigs and terminals of healthy trees before attacking the main stem of weakened individuals (Fig. 2.4f). During this "maturation feeding," trees may be infected with pathogenic fungi or nematodes carried by the beetles (e.g., Dutch elm disease transmitted by *Scolytus* bark beetles or pine-wood nematodes carried by longhorned beetles of the genus *Monochamus*; see Chapter 3). Trees weakened by pathogenic infections may then serve as a breeding substrate for the beetle vectors.

A large number of sap-sucking aphids and scales (Homoptera) also feed on the boles, branches, and twigs of forest trees, but rarely do they cause economically significant damage. Notable exceptions are the balsam woolly adelgid and beech scale, which inject salivary toxins into their hosts, and a leafhopper, which transmits the phloem necrosis virus of elm.

2.5. INSECTS FEEDING ON ROOTS

The roots of forest trees are fed upon by wireworms (Fig. 2.5a), white grubs (Fig. 2.5b), bark beetles, roundheaded and flatheaded borers, root weevils (all

TABLE 2.1

Examples of Bark Beetle Aggressiveness, or Their Ability to Overcome Healthy Trees, Which Depends on Their Tolerance of Host Defensive Chemicals, the Pathogenicity of Their Fungal Symbionts, and/or the Attractiveness of Their Aggregation Pheromones

Beetle species	Tree species	Tolerance of defense chemicals	Fungus species and pathogenicity	Attractiveness of pheromones	Aggressiveness
Mountain pine beetle *Dendroctonus ponderosae*	Pine *Pinus* spp.	Very tolerant of host terpenes	*Ceratocystis clavigera* Moderate–high	Moderate–high	High
Great spruce beetle *Dendroctonus micans*	Spruce *Picea* spp.	Very tolerant of host resin	No fungi	No adult pheromones	Low–moderate
Spruce bark beetle *Ips typographus*	Spruce *Picea* spp.	Moderate tolerance of host terpenes	*Ceratocystis polonica* High	High	High
Pine engraver *Ips pini*	Pine *Pinus* spp.	Low tolerance of host terpenes	*Ceratocystis* spp. Moderate	High	Low
Elm bark beetle *Scolytus multistriatus*	Elm *Ulmus* spp.	Probably low tolerance	*Ceratocystis ulmi* Very high in unadapted species	High	High in exotic situations
Fir engraver *Scolytus ventralis*	Fir *Abies* spp.	Intolerant of host terpenes	*Trichosporium symbioticum* Moderate	Moderate	Low

FIGURE 2.5. Feeding by root insects is most significant on the delicate root systems of newly planted seedlings. (a) A wireworm, the larva of a click beetle (Coleoptera, Elateridae). (b) A white grub, the larvae of a scarab beetle (Coleoptera, Scarabaeidae). (c) An adult root weevil (Coleoptera, Curculionidae: *Hylobius* sp.) feeding on the bark of a seedling.

beetles in the order Coleoptera), cutworms (Lepidoptera), carpenter ants (Hymenoptera), root aphids (Homoptera), and root maggots (Diptera). In general these insects are not a serious problem in established stands that have well developed and extensive root systems. However, they may provide entry for root-decaying fungi. In nurseries and newly stocked plantations, on the other hand, root insects can cause severe problems because roots are small and fragile. This problem is intensified when soil cultivation or site preparation removes the other vegetation on the site. With most of their food destroyed, root insects are then forced to feed on the roots of the planted stock.

One of the most severe problems of reforestation in eastern North America and Northern Europe is due to root collar weevils (Coleoptera), which breed in the stumps of freshly cut trees. The damage is caused by the adult *Hylobius* weevils when they emerge from stumps and then feed on the bark of newly planted seedlings (Fig. 2.5c). Feeding by large numbers of weevils in clearcut areas can frustrate reforestation plans.

In this chapter we have looked at the various types of pests encountered by the forest manager, albeit superficially. Those requiring more detailed information on specific insects, or those needing to identify a particular problem associated with a particular tree species, should consult regional reference books such as *Eastern Forest Insects,* cited at the end of this chapter. With this brief introduction to the insects and the damage they cause, we now proceed to the heart of the forest pest problem, the study of interrelationships among insect herbivores, their host plants, and the organisms that prey on them, as well as the causes of insect pest outbreaks.

REFERENCES AND SELECTED READINGS

Baker, W. L., 1972, Eastern forest insects, U.S. Forest Service, Miscellaneous Publication No. 1175.

Evans, D., 1982, Pine shoot insects common in British Columbia, Environment Canada, Canadian Forestry Service BC-X-233.

Furniss, R. L., and Carolin, V. M., 1977, Western forest insects, U.S. Forest Service Miscellaneous Publication No. 273.

Hedlin, A. G., Yates, H. D., III, Tovar, D. C., Ebel, B. H., Koorbas, T. W., and Merkel, E. P., 1980, Cone and seed insects of North American conifers, Environment Canada, Canadian Forestry Service, Ottawa.

Johnson, W. T., and Lyon, H. H., 1976, *Insects that Feed on Trees and Shrubs,* Cornell University Press, Ithaca.

MacAloney, H. J., and Ewan, H. G., 1964, Identification of hardwood insects by type of tree injury, U.S. Forest Service Research Paper LS-11.

Roques, A., 1983, *Les insectes ravageurs des cônes et graines de conifères en France,* Institut National de la Recherche Agronomique, Paris.

Rose, A. H., and Lindquist, O. H., 1973, Insects of eastern pines, Environment Canada, Canadian Forest Service Publication No. 1313.

Ruth, D. S., Miller, G. E., and Sutherland, J. R., 1982, A guide to common insect pests and diseases in spruce seed orchards in British Columbia, Environment Canada, Canadian Forestry Service BC-X-231.

Wilson, L. F., 1977, A guide to insect injury to conifers in the Lake States, U.S. Forest Service, Agricultural Handbook No. 501.

Wood, S. L. 1963, A revision of the bark beetle genus *Dendroctonus* Erickson (Coleoptera:Scolytidae), *Great Basin Natur.* **23**:1–117.

PART I: INSECT EXERCISES

Exercises are designed to provide the student with experience in searching the literature for information, analyzing population patterns and behavior, and formulating pest-management strategies. In addition, students can develop their own special project on a specific forest pest. Examples can be selected from Chapter 11, Section 11.4, or from the literature. If you plan to perform an analysis on a particular insect problem, select your insect and do the following (use drawings where possible):

I.1. Describe the biology, behavior, and anatomical features of the insect.

I.2. Discuss the strengths and weaknesses of the insect, particularly those biological features that will make it difficult to deal with or that can be used to control it.

I.3. Describe the kind of damage caused by the insect and its significance in multiple-use forestry.

PART II

ECOLOGY

"the field of forest entomology must in its very nature rest upon an ecological foundation"

S. A. Graham, 1965, *Annu. Rev.*

Entomol., Vol. 1.

Drawing by Paul A. Hengel, from *Western Wildlands*, Volume 9, Spring 1983.

CHAPTER 3

INSECTS IN THE FOREST ENVIRONMENT

We have seen that insects are a diverse and highly successful group of organisms that have evolved sophisticated morphological and behavioral adaptations for feeding on all parts of forest trees and for surviving the hazards of their environments. Trees, on the other hand, have not remained passive spectators to these assaults, having evolved equally complex defenses against insect attack. In order to understand the interactions between trees and insects, therefore, we need to spend some time studying the defensive strategies of forest trees.

3.1. TREE DEFENSES AGAINST INSECTS

For the purposes of this discussion, we will classify forest tree defenses into two basic types depending on whether they are evoked before or after insect attack. The first strategy, which we will call *preformed static defense,* is analogous to a fortified city because it employs physical or chemical barriers that impair insect feeding or digestion. For instance, some plant leaves have thorns along their edges, where many defoliators feed, whereas other plant leaves have long hairs on their undersides that deter aphids and scales. Besides physical structures, plants manufacture a vast array of chemical compounds for use as static defenses. For example, most conifers possess resin ducts, blisters, or cavities that secrete their contents into wounded areas. These resinous materials contain terpenes, phenols, tannins, and other substances that are toxic or inhibitory or that interfere with the digestive processes of insects. Interestingly, some tree species, particularly the true firs, incorporate substances into their tissues that mimic the juvenile hormone of insects. This hormone prevents the immature insect from developing into the adult (Chapter 1), and so the tree may interrupt, or at least delay, insect metamorphosis.

Certain tree species seem to have been more successful than others at developing static defense systems. Yew trees, for example, must have evolved some very effective chemicals because few insects feed on members of this genus (*Taxus*). In general, long-lived trees such as redwoods, cedars, and yews seem to possess more effective static defenses and have fewer insect enemies than do shorter-lived pines, spruces, and firs.

The second major defensive strategy is to intoxify, repel, or interfere with the reproduction or development of the insect after the attack has been initiated. These *induced dynamic defenses* are analogous to the military strategy of holding troops and weapons in reserve to attack the enemy after the invasion occurs. Dynamic responses to insect attack are sometimes called *dynamic wound responses* or *hypersensitive reactions*. Dynamic defenses are particularly important when the consequences of successful attack are severe. For example, successful attack by bark beetles usually means death for the tree, making it critical for the plant to respond quickly and strongly at the point of invasion. Conifers react to bark beetle attack by mobilizing defensive chemicals at the attack site; if the accumulation of these chemicals is rapid enough, the beetle is repelled or killed. It is interesting that dynamic responses are also effective against pathogenic microorganisms such as fungi, bacteria, and nematodes. Obviously this is an important requirement in the defense against bark beetles that carry pathogenic fungi into their feeding tunnels.

The hypersensitive reaction has been observed in many plant species and is, perhaps, a universal defense against any type of wounding or infection; i.e., it is a generalized defensive response. Dynamic wound responses are controlled graded reactions that proceed in advance of the pathogenic invasion, laying down a barrier of dead resin-impregnated tissue that lacks the nutrients essential for pathogen growth and development (Fig. 3.1). In other words. it acts as a ''scorched earth strategy'' in much the same way that the Russians burned their fields in front of Napoleon's advancing armies. In general, dynamic responses are most effective against organisms that invade the living tissues of the tree, such as bark beetles, roundheaded and flatheaded borers, tip and shoot borers, aphids and scales, and pathogenic fungi, bacteria, and nematodes. However, plants can also respond actively to the feeding of insects that browse on their foliage. The needles of larch trees, for example, are smaller and tougher and contain lower concentrations of nitrogen the year after feeding by the larch budmoth, while leaves of birch, oak, alder, and willows contain higher concentrations of phenolics, tannins, and other defensive chemicals after adjacent leaves have been damaged by defoliators.

The dynamic response of plants to defoliation is much less intense and concentrated than that to boring insects; also, the chemicals are not normally toxic but act, rather, to slow down growth and development or to reduce fecundity. Plants responding to defoliation may produce tannins that bind with proteins and carbohydrates, making them indigestible, or proteolytic enzyme inhibitors that interfere with the digestive process. Plants may also tie up their

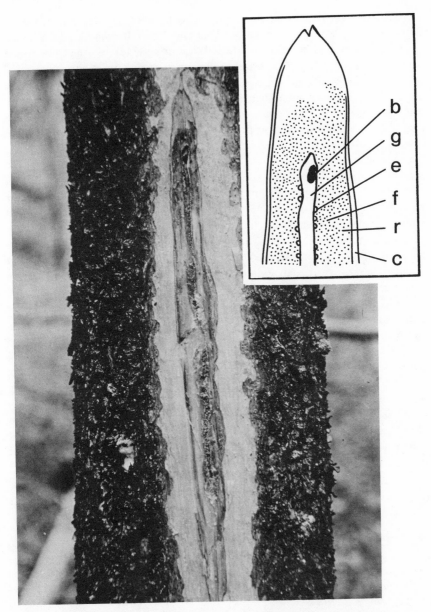

FIGURE 3.1. Bark removed from a lodgepole pine to show a stabilized dynamic wound response to a mountain pine beetle attack. The beetle gallery and surrounding phloem tissue are soaked with resinous materials, and the female beetle and her eggs have been intoxified by the resin. The drawing at the top right illustrates the anatomy of the wound response in more detail: insect gallery (g) containing dead beetle (b) and eggs (e); fungus-infected tissue soaked in resin (f); fungus-free resin-soaked tissue (r); wound periderm and callus tissue formed around the perimeter of the stabilized lesion (c), which will eventually grow over and compartmentalize the wound.

nitrogen, making it unavailable to the feeding insect, or may produce more fibrous needles, which are harder to devour and digest. In addition, the dynamic responses to defoliation are often diffused, occurring in leaves or needles distant from the point of injury. In contrast to the hypersensitive reaction to borers and pathogens, therefore, the defense against defoliators is less intense but more extensive. This seems a reasonable strategy against insects that do not normally cause severe injury to the plant; i.e., plants, especially deciduous species, can lose a considerable proportion of their foliage without suffering irreversible physiological stress.

The successful utilization of dynamic defenses rests on the ability of the plant to *recognize* the presence of the invading organism. It is generally thought that detection is made possible by chemical *elicitors* secreted by the pathogen or produced as a result of feeding activities (Fig. 3.2). These elicitors set in motion a chain of metabolic activities that result in the synthesis of defensive chemicals. In order to "learn" which substances to utilize as elicitors, however, the plant must have been exposed to the pathogen in the evolutionary past. Only through the process of adaptation over long periods of time can a plant develop the genetic ability to sense the presence of pathogenic organisms. Consequently, it should not surprise us that plants that have not had the chance to adapt to their enemies are in serious danger. Thus, American elms are being eliminated by the Dutch elm disease fungus, chestnuts by a blight fungus, western white pines by a rust fungus, firs by a woolly adelgid, and beech by a bark disease associated with scale insects, all of which have been introduced from other countries. Similarly, Japanese pines are being decimated by the pine-wood nematode, apparently introduced from America, Monterey pine plantations in Australia are ravaged by a European woodwasp, and European elms are being destroyed by a new strain of the Dutch elm disease fungus that seems to have evolved from the mixing of American and Far Eastern varieties. These disasters should always remind us that

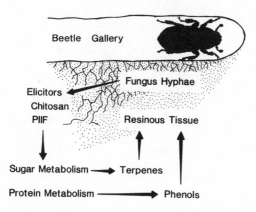

FIGURE 3.2. Schematic illustration of the dynamic wound response of pine trees to invasion by bark beetles and their associated fungi. Growth of fungal hyphae into the living tissue-release elicitors that stimulate the synthesis of terpenes and phenols (and perhaps other substances) through the sugar and protein metabolic pathways; these create a resinous matrix in the phloem and xylem tissues that impede and eventually stop fungus penetration.

trees require evolutionary time, possibly thousands of years, to develop the sensory capabilities to recognize and defend against their enemies.

The induced dynamic defenses of plants are similar, in some ways, to the immunological systems of vertebrates. The induction of a generalized defensive response in cells close to (bark beetles) or distant (defoliators) from the point of injury rests on the plants ability to recognize the specific pathogen or insect. These similarities suggest that plants may be able to acquire immunity to insect/pathogen attack after an initial sublethal infection. Although acquired immunity has been demonstrated in some agricultural plants, it is not known whether forest trees have the same capacity. If they do, it opens up a whole new concept in forest protection—the vaccination and immunization of trees against insects and diseases.

Another fascinating aspect of dynamic plant defense recently came to light when researchers discovered elevated levels of defensive chemicals, not only in the leaves of willow trees defoliated by tent caterpillars, but in adjacent unattacked trees as well. Carefully controlled laboratory experiments seem to confirm that trees can communicate the presence of feeding insects to other nearby trees. Although the mode of communication is unknown, plants have long been known to produce the highly volatile compound ethylene in wounded tissues. One is tempted to speculate that ethylene acts as the messenger that informs nearby tissues to begin defensive metabolism and that the diffusion of ethylene into the atmosphere may trigger similar processes in nearby trees.

Although chemical defense is an effective strategy against herbivorous insects, there is one major problem—the chemicals are often large molecules requiring considerable amounts of energy for synthesis. Thus, plants deficient in energy reserves (starches and sugars) may be unable to produce sufficient quantities of the chemicals to defend themselves effectively. Under these conditions they may be overwhelmed by bark beetles, or defoliator populations may grow rapidly because the chemicals that reduce their growth and reproduction are absent.

The energy reserves of plants depend on many factors, including light (for photosynthesis), soil nutrients and water, and plant aging. When trees are subjected to stress from light, nutrient, or water deficiencies or from disease or old age, they become more vulnerable to insect attack. For example, trees under severe moisture stress have weak hypersensitive reactions and are easily overcome by bark beetles, whereas defoliator populations increase more rapidly on stressed trees because higher levels of usable nitrogen and lower levels of defensive chemicals are present in their foliage. In general, then, the reproduction and survival of insect herbivores are much higher on, or in, trees that have been weakened by various stress factors; such trees will be more severely damaged than their healthy neighbors.

Complex chemical defenses cannot insure the plant immunity against insect attack. Insects, with their short generation spans, high reproductive potentials, and genetic variability, can evolve offensive strategies at much faster rates than trees are able to evolve new defensive adaptations. Some insects even use the plant's defensive chemicals for their own purposes. For example, bark beetles use plant terpenes to manufacture aggregating pheromones, and some sawflies sequester terpenes for use in their own defense. Many insects also possess intricate mechanisms for detoxifying plant chemicals.

Static defenses seem to be most vulnerable to counteroffensive adaptations because insects that rely on a particular plant for food are continuously exposed to their action. This continuous selection pressure leads to counteradapted genotypes in much the same way that insects become resistant to insecticides. Thus, although static defense may be an effective strategy against generalist herbivores (those that feed on several or many plant species), they may be less effective against the specialists that are adapted to feed on a particular plant.

Counteroffensive adaptations against dynamic defenses are not as likely to evolve in herbivore populations because weak trees, incapable of responding, are always present in the natural forest. Insects feeding on these weak trees will have higher survival and reproductive rates because defensive chemicals will be reduced or lacking. By contrast, insects feeding on healthy trees will be killed or will have much lower reproductive potentials. As a result, genotypes that attack weakened trees will survive to dilute the gene pool, and the evolution of counteroffensive adaptations will be slowed or halted.

Our understanding of the defensive physiology of forest trees and of the offensive behavior of insects is still fragmentary. As we continue to study trees and insects as dynamic entities in the drama of attack and defense, more surprising and fascinating aspects of their interaction will undoubtedly come to light. These discoveries will increase our abilities to manage insect pest populations by improving the defensive capabilities of forest trees.

3.2. INSECTS AND FOREST STABILITY

The concept of forest stability tends to be clouded by the use of mystical terms such as "the balance of nature" and "homeostasis." But in its strict sense there is nothing magical about the idea of stability. A stable ecosystem is defined as one that remains close to an equilibrium state, in terms of the numbers of species and their relative abundances, for long periods of time, and that tends to return to that equilibrium if disturbed. If we accept this definition, we can apply some important rules from what is called *general systems theory* to help us understand how stable forest systems arise.

The first rule is that stability can only be imposed through the action of *feedback processes*. Feedback is an elementary concept that arises from consideration of systems in which an initial stimulus is fed back to its origin to create a *feedback loop*. Thus, if I speak to you and you say something back to me that causes me to speak again, we have taken part in a feedback loop. There are only two kinds of feedback, positive or negative, but their effects are drastically different. *Positive feedback* results when an initial stimulus passes through a feedback loop, eventually causing an increase (+) in the original stimulus. Positive feedback is characteristic of unstable "vicious circles" like the arms race, the inflation spiral, and the population explosion. *Negative feedback*, on the other hand, is inherently stabilizing and results when an initial stimulus passes through a feedback loop to eventually cause a decrease (−) in that stimulus. Negative feedback therefore tends to produce equilibrium or steady-state conditions as exemplified by thermostats, governors, autopilots, and other control devices.

Having dispensed with these elementary definitions, let us return to the problem of stable forest ecosystems. Perhaps we can illuminate this problem best by considering an unstable system, such as the pine-wilt disease currently ravaging Japanese forests. The causal agent of pine-wilt disease is a nematode that infects the resin ducts and feeds on the living duct epithelium, parenchyma cells, and possibly the phloem tissues. The symptoms of the disease are a rapid decline and eventual death of the infected tree, the needles turning yellow and later a bright red as the tree dies. Nematodes are transmitted from dead trees to living ones by a roundheaded borer (Cerambycidae) which breeds in the dying plant. As the adult beetles emerge from dead pines, the nematodes conceal themselves in the thoracic spiracles and under the forewings. Newly emerged beetles then feed on the elongating shoots of uninfested pines. On this new host, the nematodes abandon their hiding places and enter the resin ducts through feeding wounds made by the insect. The feedback structure of this system is illustrated in Fig. 3.3. It follows that an initial nematode infection creates breeding material for the cerambycid, causing an increases (+) in the number of beetles, which then increase (+) the number of new nematode infections (i.e., more beetles infect more trees). The rules of feedback specify that the sign of a loop is determined by multiplying the signs of all the interaction pathways; in this case, therefore, we have an unstable positive feedback loop because $(+) \times (+) \times (+) = (+)$ (Fig. 3.3). The continued operation of this positive loop will eventually lead to the extinction of the pines, the nematode, or both, provided that no changes occur in its structure.

The nematode–pine interaction is an example of an unadapted host–parasite system (the nematode was probably introduced, accidentally, from North America). As we discussed in the previous section, evolutionary time is required for the tree to evolve mechanisms for recognizing the new pathogen. We might suspect that a few Japanese pines have this intrinsic ability, and that these individuals will survive the onslaught of the pine-wood nematode to give rise to a new generation of resistant trees.

The pine-wood nematode is an example of a problem which has plagued man throughout his history as he accidentally introduces new pests into the forests of the world, or moves tree species to new places. We might just as well

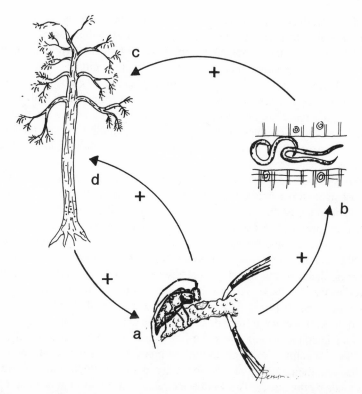

FIGURE 3.3. Unstable system dominated by positive feedback. (a) Cerambycid beetles (*Mono-chamus* sp.) emerging from dead pines feed on the twigs of healthy trees, and the nematodes (*Bursaphelenchus xylophilus*), which are carried under the elytra and in thoracic spiracles of the beetle, gain entry through the open wound. (b) Nematodes reproduce in the resin ducts of the tree and move through the xylem *via* these ducts. (c) Infected trees become chlorotic and die in about 3 months due to the destruction of the resin duct epithelium by the nematode and its associated bacteria or by the action of toxins excreted by these organisms. (d) Dead or dying trees provide a breeding substrate for the cerambycid. Hence, an increase in the number of infected pines (+ interaction) increases the number of cerambycid beetles (+ interaction), thereby increasing the number of nematode infections (+ interaction). The overall effect of a feedback loop is the product of its interactions: $(+)\times(+)\times(+) = (+)$ feedback.

have used the Dutch elm disease, white pine blister rust, chestnut blight, or the balsam woolly adelgid as examples of unstable positive feedback systems.

Let us now turn our attention to more normal conditions, where the insect and its food plant have coevolved together over extended periods of time. As an example we will look at bark beetles and their native hosts, mainly because these insects also carry pathogens into the tree. In this case, however, healthy trees are able to recognize the presence of the pathogen and, by mobilizing large quantities of terpenes and phenolics at the site of infection (hypersensitive responses), usually succeed in sealing off the insect and fungus in a lesion of dead resinous tissue (Figs. 3.1 and 3.2). It is only the weak and unhealthy trees which are unable to defend themselves because they lack the necessary energy. If we assume that the number of weakened trees in a stand is directly proportional to stand density, then we can construct the feedback structure shown in figure 3.4;

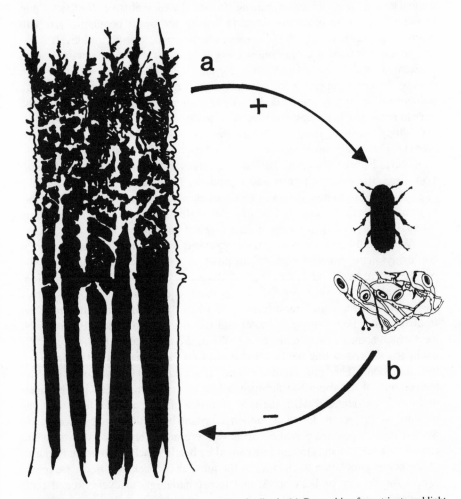

FIGURE 3.4. Stable system dominated by negative feedback. (a) Competition for nutrients and light in dense stands reduces tree vigor and leads to an increase in the population of bark beetles and their associated fungi. (b) Thinning attributable to bark beetle-caused mortality reduces competition and recycles nutrients to the more vigorous stems. Thus, an increase in stand density leads to an increase in the abundance of beetles and fungus (+ interaction); stand density then decreases because weak trees are killed (− interaction). The overall effect is a stabilizing negative feedback loop: $(+) \times (-) = (-)$ feedback.

this assumption is fairly reasonable because competition between trees for light, soil nutrients, and water becomes more intense as stand density increases, as does the spread of root-rot fungi and other debilitating agents. We can see from this diagram that increasing stand density causes the bark beetle population to increase (+), because more weak trees are available for colonization, and this results in a decrease in stand density as the weak trees are killed (−). The total feedback structure, therefore, creates a stabilizing negative feedback loop (+)×(−) = (−). In other words, stand density and beetle population size will automatically adjust, with time, to some characteristic equilibrium level.

In the preceeding discussion we assumed that the environment remained constant so that the only factor affecting the health of the trees was their own density. In a changing environment, however, we would expect other factors to influence plant vigor. Suppose, for example, that a drought weakened all the trees in a stand. The beetles would respond to this event by increasing in numbers and killing many of the trees before they recovered from the effects of the drought. Eventually the drought will end and the beetle population will decline to a low level because there will be few weak trees in this sparsely stocked stand. With time, the trees will grow and reproduce, their density will rise, and the original equilibrium between trees and beetles will be attained once more. This sequence of events is characteristic of stable systems that have been disturbed by external factors and then return to their original equilibrium states.

We should be aware that systems regulated by negative feedback loops do not always have constant equilibrium positions. Under certain conditions the populations of trees and insects may oscillate or cycle around their equilibrium levels. This occurs when populations approach equilibrium too rapidly so that they overshoot the equilibrium point, and when time delays are present in the feedback loop. For example, a heater and thermostat form a negative feedback loop which controls room temperature. When the temperature of the room falls below the desired setting the thermostat turns on the heater and hot air is blown into the room. This process takes time, however, and in the meantime the temperature of the room has dropped below its equilibrium setting. The same problem is encountered when the heat is turned off so that, in effect, the temperature of the room does not remain constant but rather cycles around the desired equilibrium temperature. Because there are likely to be delays in the response of forest stands to thinning caused by beetle attacks, and in the response of the beetle population to changes in the number of weakened trees, we might expect the density of both beetle and tree populations to cycle around their equilibrium levels.

It should also be noted that some natural systems contain both negative and positive feedback loops and, in these cases, we should see stable conditions at certain times and unstable situations at others. For example, certain bark beetles are able to overcome the resistance of trees by rapid "mass attack" mediated by attractive aggregating pheromones (Chapter 2). When beetle populations are high, aggregation is so rapid that even healthy trees can be overwhelmed. This gives rise to an unstable positive feedback loop; that is, an increase in beetle population density enables them to attack more trees (+), which then increases the beetle population (+). When beetle populations are sparse, however, aggregation occurs very slowly so that only weak trees can be successfully attacked (i.e., the negative feedback loop of Fig. 3.4 now dominates the system). The transition from stable, sparse populations to unstable dense populations is often triggered by environmental disturbances such as droughts or other stress factors that temporarily increase the beetle population and initiate the positive feedback process. Systems that remain in a stable state for much of the time but that are ocassionally destabilized by environmental disturbances have been called *metastable* to draw attention to their bounded or limited stability conditions (*meta* = bounded). These interesting topics will be discussed more formally in Chapter 4.

We have seen that adaptive evolution between trees and insects is essential for the development of stable interactions, and that stability is enforced through the action of negative feedback loops. These same concepts can be applied to forest communities composed of many interacting species. For example, Figure 3.5 shows the interactions in a hypothetical community consisting of three *trophic levels*: the producers that manufacture carbohydrates from raw materials and the sun's energy (e.g., trees), herbivores that feed on the producers (e.g., phytophagous insects), and a carnivore feeding on the herbivores (e.g., a predaceous insect). A structure such as this is termed a *food web*, and it usually has a triangular shape because energy is lost in transference from one trophic level to another; i.e., some of the material flowing from plants to herbivores is expended in herbivore respiration so that it does not contribute to herbivore biomass. Although herbivore biomass will therefore be less than that of the producers, the number of species present in the higher trophic level may or may not be less.

It is important to realize that competitive interactions within each trophic level give rise to unstable positive feedback loops. For example, competition between tree species P_1 and P_2 (Fig. 3.5) results in P_1 having a negative effect on the reproduction and survival of P_2 and *vice versa*, producing a positive feedback loop $[(-)\times(-) = (+)]$. These unstable competitive interactions are the

FIGURE 3.5. Food web consisting of three producers (plants), two herbivores, and one carnivore. Note that all competitive feedback loops within each trophic level are positive $[(-)\times(-) = (+)]$, that all feeding loops between trophic levels are negative $[(+)\times(-) = (-)]$, and that longer loops linking three or more species may be positive or negative.

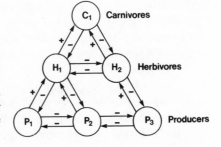

driving force of forest succession because one species tends to be driven out from an area and replaced by the other. In contrast, interactions between trophic levels create stabilizing negative feedback loops; e.g., the interaction between producer P_1 and herbivore H_1 gives $(+) \times (-) = (-)$.

In addition to the competitive and trophic loops linking two species, complex ecosystems contain longer feedback loops involving several species. For example, we can trace the loop $H_1 \xrightarrow{-} H_2 \xrightarrow{-} P_3 \xrightarrow{-} P_2 \xrightarrow{+} H_1$ (see Fig. 3.5), which is composed of four interactions $(-) \times (-) \times (-) \times (+) = (-)$; i.e., it is a stabilizing negative feedback loop. (*Note*: the student should work out the stability properties of other loops in Fig. 3.5 involving three to five species.) The overall stability of a complex community is determined by the combined effect of all its feedback loops, being stable if negative feedback dominates and unstable otherwise. In most cases, however, we would expect natural ecosystems to be relatively stable, at least in the long term, because their interactions are the result of evolutionary adaptations between the various organisms and because unstable structures would have led to species extinctions.

Insects can play an important role in the maintenance of stable forest communities. For example, should one of the tree species become very abundant, say P_2 in Fig. 3.5, herbivore H_1 may feed more heavily on this species, giving its competitor P_1 an advantage. Similarly, should herbivore H_1 become very abundant, the carnivore C_1 may switch its feeding to this species, forcing it back toward its original level. Responses to changes in the abundance of various species often reverberate through a food web and tend to regulate the composition and relative abundance of the various species at levels characterizing that particular community. In other words, they tend to stabilize the ecosystem.

3.3. INSECTS AND FOREST SUCCESSION

Climax forests are usually considered the most stable associations of tree species that can grow on a particular site. Climax communities are usually composed of long-lived, shade-tolerant, and insect-resistant tree species (Table 3.1). However, when climax forests are destroyed by cataclysmic events, such as fires or clearcutting, the site is usually occupied by *pioneer* species, which grow best on bare mineral soil and under exposed conditions. These pioneer species tend to be intolerant of shaded conditions and therefore cannot reproduce under their own canopy. Thus, shade-tolerant climax species become established in the understory and eventually dominate the site as the pioneer species die off. This process of gradual change toward a stable climax community is known as ecological *succession*.

Insects often play a significant role in forest succession, either speeding up or slowing down the rate at which succession proceeds. For example, many intolerant species are susceptible to attack by tree-killing bark beetles (Table 3.1). By removing these pioneer species before they would normally die of old

TABLE 3.1
The Number of Species of Foliage-Feeding Caterpillars
and Bark-Boring Beetles that Attack Selected Western Conifers
Compared with the Respective Shade Tolerance and Longevity of the Trees

Tree species	Common name	Tolerance[a]	Longevity[b]	Number of species[c] Browsers	Bark beetles
Sequoia sempervirens	Redwood	2	2	2	1
Libocedrus decurrens	Incense cedar	2	3	6	2
Thuja plicata	Western red cedar	1	2	11	3
Abies lasiocarpa	Subalpine fir	1	4	6	9
Larix occidentalis	Western larch	5	2	12	4
Pinus monticola	Western white pine	3	3	6	12
Abies concolor	White fir	2	4	18	10
Picea sitchensis	Sitka spruce	2	2	14	14
Picea engelmannii	Englemann spruce	2	3	12	16
Tsuga heterophylla	Western hemlock	1	3	22	7
Pinus contorta	Lodgepole pine	4	4	33	25
Pinus ponderosa	Ponderosa pine	4	4	42	23
Pseudotsuga menziesii	Douglas-fir	3	2	49	24

[a]Relative tolerance to shade: 1 = very tolerant; 5 = very intolerant.
[b]Relative longevity: 1 = very long lived; 4 = relatively short lived.
[c]Larvae of foliage-feeding Lepidoptera and bark beetles. (Counted from Furniss and Carolin, 1977.)

age, the insects hasten the transition to the climax forest (Fig. 3.6a). In some cases, however, insects may be instrumental in retarding or halting the successional trend. For example, trees killed by bark beetles also produce a tremendous quantity of fuel, thereby increasing the probability that a forest fire will kill the climax species in the understory and prepare the way for the next generation of pioneers (Fig. 3.6b). Indeed, it has been suggested that some pines have co-adapted to their bark beetle enemies to such an extent that they become susceptible to attack at an age that maximizes their offspring's chance of successfully reoccupying the site.

We should realize that there are exceptions to the above generalities. Some intolerant pioneer species seem to be unusually resistant to insect attack (e.g., western larch, Table 3.1), whereas other fairly tolerant and long-lived species seem to have many insects feeding on them (e.g., Douglas-fir, Table 3.1). The strategies of forest trees involve tradeoffs among rapid growth, reproduction, defensive metabolism, and other physiological processes and morphological structures. All these processes are dependent on energy, and energy put into growth, maintenance, and reproduction cannot be used for defense against insects. Energy for defensive purposes can be conserved by increasing photosynthetic efficiency (tolerant species), reducing growth and reproductive rates,

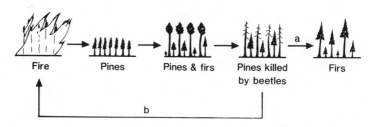

Fire   Pines   Pines & firs   Pines killed   Firs
                                         by beetles

b

FIGURE 3.6. Succession of pines by fir. (a) hastened succession caused by bark beetles killing pines, and (b) retarded succession when forest fires start in beetle killed pines, preparing the way for the next pine generation.

and/or reducing maintenance costs (deciduous species). Different tree species have evolved different strategies for conserving and allocating energy for defensive purposes.

3.4. INSECTS AND FOREST PRODUCTIVITY

Earlier in this chapter we saw that stress resulting from either aging, competition for water, nutrients or light, or from root or foliage damage by pathogens or insects can reduce the effectiveness of plant defensive systems. Because of this, insects will usually be more successful at attacking and will reproduce more offspring on trees that are under the greatest stress. When these trees are killed, their nutrients will be recycled back to the healthy survivors. This chain of logic is important because it implies that insects regulate, in a loose sense, the average productivity of forest trees. In other words, the actions of insects that cause nutrients to be cycled from weak to healthy trees tend to maximize the rate of biomass accumulation or the productivity of the average tree. This concept is illustrated schematically in Fig. 3.7. Here we see that insects are involved in two negative feedback loops that influence tree growth or vigor. In the first loop (a $\rightarrow$ d $\rightarrow$ c $\rightarrow$ a), the insects affect growth rates by thinning the stand and reducing the intensity of competition for light, water, and nutrients. In the second loop (a $\overset{+}{\rightarrow}$ b $\overset{+}{\rightarrow}$ c $\overset{-}{\rightarrow}$ a), they affect growth rates by recycling nutrients in the form of excreta, dead insects, and dead plant parts. As all the loops in this system are negative, the interaction network creates a stable negative-feedback system in which insect numbers, forest biomass density, soil nutrients, and tree growth attain characteristic equilibrium values. In fact, we might expect these values to cycle around their equilibria because some of the processes involved in the feedback loops are rather slow acting, e.g., the decay and breakdown of plant material into its raw elements and the growth response of the forest after thinning. As we have noted previously, time delays in the action of negative feedback loops can create cycles of abundance (see Fig. 3.8).

Although insects may regulate average plant growth close to its maximum

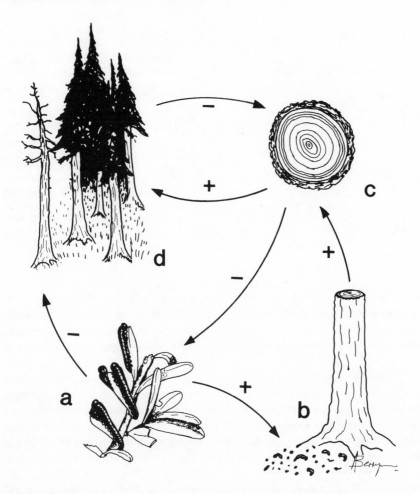

FIGURE 3.7. Interactions among (a) insect population, (b) soil nutrients, (c) tree growth and vigor, and (d) stand density. Note negative feedback loops created by stand thinning [a $\xrightarrow{+}$ d $\xrightarrow{+}$ c $\xrightarrow{+}$ a = $(-)\times(-)\times(-) = (-)$] and nutrient recycling [a $\xrightarrow{-}$ b $\xrightarrow{-}$ c $\xrightarrow{-}$ a = $(+)\times(+)\times() = (-)$].

level for a particular site, this natural process often conflicts with man's short-range management goals. As we can see in Fig. 3.8, insect damage often occurs close to the time when the forest reaches its maximum biomass density or when the standing crop has reached its greatest volume. From the manager's standpoint, the insect causes severe damage when the forest has attained its maximum value in terms of salable timber volume. The lesson to be learned, however, is fairly straightforward: In order to use timber volume that would be destroyed by insects, the manager must harvest the trees before they become susceptible to insect attack (i.e., at the end of the period of maximum growth). An even more

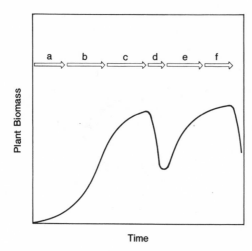

FIGURE 3.8. Growth cycle of a forest stand and the role of insects in thinning the stand and nutrient recycling. (a) Establishment phase: water, nutrients, and light not limiting. (b) Maximum growth phase: water, nutrients, and light not limiting. (c) Maximum biomass phase: water, nutrients, and/or light limit growth. (d) Biomass reduction phase: insects and pathogens thin weakened trees from the stand. (e) Growth recovery phase (second cycle): nutrients, water, and light no longer limiting. (f) Maximum biomass phase (second cycle); return to (c).

effective strategy would be to thin the stand when its growth rate begins to decline. In this way, maximum growth rates can be maintained on the residual trees, and financial rewards may be reaped from the sale of timber removed during the thinning operations before it is destroyed or degraded by insects and disease.

Before leaving this general ecological discussion, we should mention some of the other roles played by insects in forest ecosystems. For instance, some insects such as wasps and bees are important pollinators of flowering trees and shrubs, ensuring cross-fertilization in these species. Other insects (e.g., ants, termites, and wood-boring beetles) bore into the wood of dead trees, increasing the invasion rate of wood-decaying pathogens and the rate of wood decay and nutrient turnover. This results in decreased fuel loads and thereby lowers the likelihood of forest fires. Other soil-inhabiting insects such as collembolans, thysanurans, and certain beetles and flies speed nutrient recycling by feeding on decaying organic matter, fungi, and so forth. Lastly, numerous insect species act as predators or parasites of herbivorous pests. These natural enemies help control pest populations and often prevent them from reaching outbreak levels (see Chapter 5). In addition to the insects, many insectivorous birds and mammals live in the forest environment; these may also play important roles in suppressing pest populations. These allies of ours should always be kept in mind as we manipulate forest environments by cutting, fertilizing, or spraying insecticides and herbicides, for, as we have seen, effects on one component of a complex forest food web may reverberate through the entire system and result in undesirable consequences, one of which may be outbreaks of destructive herbivorous insects. We should always remember that the vast majority of plant-feeding insects inhabiting our forests are maintained at very low densities by interactions with the defenses of their hosts or with their natural enemies and that human interference can disrupt this natural balance and create new pest problems.

REFERENCES AND SELECTED READINGS

Baker, F. S., 1950, *Principles of Silviculture,* McGraw-Hill, New York.

Baldwin, I. T., and Schultz, J. C., 1983, Rapid changes in tree leaf chemistry induced by damage: Evidence for communication between plants, *Science* **221**:277–279. (defense communication)

Berryman, A. A., 1972, Resistance of conifers to invasion by bark beetle–fungus associations, *BioScience* **22**:598–602. (induced and static defenses)

Berryman, A. A., 1981, *Population Systems: A General Introduction,* Plenum Press, New York.

Berryman, A. A., Stenseth, N. C., and Wollkind, D. J., 1984, Metastability of forest ecosystems infested by bark beetles, *Res. Pop. Ecol.* **26**:13–29. (metastable forest–insect systems)

Carter, C. I., and Nichols, J. F. A., 1985, Some resistance features of trees that influence the establishment and development of aphid colonies, *Z. Angew. Entomol.* **99**:64–67.

Cowling, E. B., and Horsfall, J. G., 1980, How plants defend themselves, in: *Plant Disease— An Advanced Treatise,* Vol. 5 (J. G. Horsfall and E. B. Cowling, eds.), Academic Press, New York. (general overview of defense against disease)

Feeny, P., 1976, Plant apparency and chemical defense, *Recent Adv. Phytochem.* **10**:1–40. (theory of plant defense)

Furniss, R. L., and Carolin, V. M., 1977, Western forest insects, U.S. Department of Agriculture Miscellaneous Publication No. 1339.

Hanover, J. W., 1975, Physiology of tree resistance to insects, *Annu. Rev. Entom.* **20**:75–95. (mainly static defenses)

Harper, J. L., 1969, The role of predation in vegetational diversity, *Brookhaven Symp. Biol.* **22**:48–62. (forest stability)

Haukioja, E., 1980, On the role of plant defenses in the fluctuation of herbivore populations, *Oikos* **35**:202–213. (delayed induced defense and cycles)

Kobayashi, F., Yamane, A., and Ikeda, T., 1984, The Japanese pine sawyer beetle as the vector of pine wilt disease, *Annu. Rev. Entomol.* **29**:115–135.

Kuć, J., 1982, Induced immunity to plant disease, *BioScience* **32**:854–860.(immunization)

Mattson, W. J., Jr., 1980, Herbivory in relation to plant nitrogen content, *Annu. Rev. Ecol. Sys.* **11**:119–161. (importance of nitrogen)

Mattson, W. J., Jr., and Addy, N. D., 1975, Phytophagous insects as regulators of forest primary production, *Science* **190**:515–522. (forest productivity)

Mullick, D. B., 1977, The non-specific nature of defense in bark and wood during wounding, insect and pathogen attack, *Recent Adv. Phytochem.* **11**:395–441 (induced defenses)

Peterman, R. M., 1978, The ecological role of mountain pine beetle in lodgepole pine forests, in: Theory and Practice of Mountain Pine Beetle Management in Lodgepole Pine Forests (A. A. Berryman, G. D. Amman, R. W. Stark, and D. L. Kibbee, eds.), pp. 16–26, Forest, Wildlife and Range Experiment Station, University of Idaho, Moscow. (coadaptation between insects and trees)

Price, P. W., 1975, *Insect Ecology,* Wiley (Interscience), New York. (community structure and coevolution)

Rhoades, D. F., 1979, Evolution of plant defense against herbivores, pp. 3–54 in: *Herbivores: Their Interaction with Secondary Plant Metabolites* (G. A. Rosenthal and D. H. Jansen, eds.), Academic Press, New York. (plant defenses)

Rhoades, D. F., 1983, Responses of alder and willow to attack by tent caterpillars and webworms: Evidence of pheromonal sensitivity of willows, in: *Plant Resistance to Insects.* (P. A. Hedlin, ed.), pp. 55–68, *ACS Symposium Series No. 208,* American Chemical Society. (defense communication)

Rhoades, D. F., 1985, Offensive–defensive interactions between herbivores and plants: Their relevance in herbivore population dynamics and ecological theory, *Am. Nat.* **125**:205–238. (plant defense and herbivore attack strategies)

Rhoades, D. F., and Cates, R. G., 1976, Towards a general theory of plant antiherbivore chemistry, *Recent Adv. Phytochem.* **10**:168–213. (theory of plant defense)

Safranyik, L. (ed.), 1985, *The Role of the Host in the Population Dynamics of Forest Insects,* Proceedings of the IUFRO Conference, Banff, Alberta, Pacific Forest Research Center, Victoria, B.C.

Schoeneweiss, D. F., 1975, Predispostion, stress, and plant disease, *Ann. Rev. Phytopathol.* **13**:193–211. (stress and defense)

Schultz, J. C., and Baldwin, I. T., 1982, Oak leaf quality declines in response to defoliation by gypsy moth larvae, *Science* **217**:149–151. (induced defense)

White, T. C. R., 1978, The importance of a relative shortage of food in animal ecology, *Oecologia (Berl.)* **33**:71–86. (particularly the role of nitrogen)

CHAPTER 4

POPULATION DYNAMICS OF FOREST INSECTS

Insects do not normally cause serious damage to forests unless their numbers increase, for some reason or another, to very high densities. The area of ecology that deals with changes in the density of organisms over time and that attempts to explain the causes of these changes is known as population dynamics. This chapter explores the concepts of population dynamics as they relate to forest insects, in order to better understand the causes of forest pest outbreaks. This basic understanding is essential for both forest managers and insect pest managers, for without it they will often find themselves groping in the dark for solutions to their pest problems and may even make decisions that compound rather than solve the problems.

4.1. PRINCIPLES OF POPULATION DYNAMICS

A population consists of a relatively large group of individuals of the same species living together in a particular setting. Thus, we may speak of the population of people in New York, of elk in Yellowstone National Park, of fish in lake Michigan, and so on. Within these broadly defined spatial boundaries we can also describe insect numbers in terms of their *density* per unit area of habitat, e.g., insects per square meter of bark, foliage, or soil, or per hectare of forest. The use of population density as a unit of measurement not only maintains the spatial dimension, because habitats provide the space in which insects live, but also relates insect numbers to the damage they cause. After all, it is the number of insects per unit area of its host that determines the degree of damage to that host. In this book, therefore, we will usually use density per unit area of habitat as the basic unit of population measurement.

4.1.1. Determinants of Population Change

Population dynamics is, in essence, the study of births, deaths, and movements of organisms, for it is these variables that cause populations to change over time. In its simplest sense, population change can be defined by the following

$$\text{Population change} \atop \text{per unit of time} = {\text{birth} \atop \text{rate}} - {\text{death} \atop \text{rate}} + {\text{immigration} \atop \text{rate}} - {\text{emigration} \atop \text{rate}}$$

This concept can be represented symbolically as follows:

$$\Delta N = B - D + I - E \qquad (4.1)$$

where ΔN is the change in population density over a given time period, measured in days, years, or insect generations, and $B, D, I,$ and E are the total numbers of births, deaths, immigrations, and emigrations occurring over the same period of time. Births, deaths, and migrations are the result of events that happen to individual organisms, for it is individuals that give birth, die, and move out of or into particular areas. Thus, we can also express population change in terms of individual rates; that is,

$$\Delta N = (b - d + i - e)N \qquad (4.2)$$

where N is the initial density of the population, b is the average birth rate per individual, and $d, i,$ and e are the rates of individual death, immigration, and emigration, all expressed within a specified time interval. It is obvious from this relationship that population density increases when individual births and immigrations exceed deaths and emigrations ($b + i > d + e$), declines when $b + i < d + e,$ and remains unchanged, or in equilibrium, when $b + i = d + e.$

These simple equations describe precisely how populations change over long periods of time (many years or generations) if and only if the birth, death, and migration rates remain constant over that period. Obviously this provision is rarely met in natural situations because the forces affecting birth, death, and migration rates change from time to time and from place to place. For this reason we need to understand how the survival, reproduction, and movement of individual organisms are influenced by the environment in which they live.

4.1.2. The Concept of Environment

For simplicity, let us consider a population as being made up of a group of identical individuals. True, each individual is unique, as are all organisms, but the members of a particular species also possess a set of genetic properties that characterize all members of that species. These specific attributes determine the resources required by individuals of the species in order to carry out their normal functions (e.g., kinds of food, nesting places), their ability to obtain resources in the face of competing organisms (food finding, attack, and competitive behaviors), and their capacity to defend themselves and escape being eaten (flight and defense physiology and behavior). Thus, these specific attributes determine how individual organisms will interact with their environment (Fig. 4.1).

An organism's environment is composed of many physical and biological elements, including land topography, soils, weather-related factors (e.g., temperature, moisture, wind), food organisms, predators, parasites, pathogens, competing and cooperating species, as well as the members of its own species. The sum of all these environmental factors determines the well-being of the individual, for they supply the resources needed for growth and reproduction and also create the risks to life. From the perspective of the individual, the environment may be favorable or unfavorable, depending on the relative abundance of these various factors (effect e in Fig. 4.1).

The concept of *environmental favorability* visualizes the environment within the context of individual needs and risks. Favorable environments provide plentiful food of good quality; a low risk of being killed by predators, parasites, and pathogens; and a good chance of finding a mate. They provide the individual

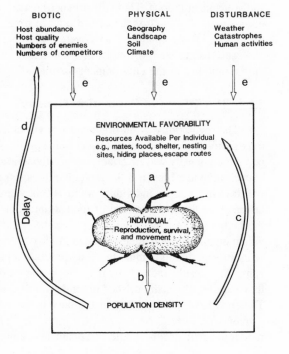

FIGURE 4.1. Conceptual view of the determinants of density changes in insect populations. The basic environment, which is subdivided into biotic, physical, and disturbance factors, together with the density of the insect population, determine the favorability of the environment for the average individual (interactions e and c). Individual rates of birth, death, and migration are determined by the resource requirements of the species together with environmental favorability (interaction a); these rates govern the growth rate of the population (interaction b). Finally, population density may affect the basic properties of the environment (interaction d) and through this the favorability of the environment for future generations (delayed feedback).

with a "good life," a high "standard of living." As a result, individuals survive and reproduce more successfully in favorable environments, and movements are generally into, rather than out of, such regions. When environments are favorable, births and immigrations usually exceed deaths and emigrations, and the population grows, i.e., $b + i > d + e$.

The favorability of an organism's environment is determined, to a great extent, by the *physical setting* in which it lives; this physical setting is sometimes called the *site*. For instance, temperature and precipitation are related to the latitude and elevation of a particular site, as well as to its position in relationship to oceans and mountain ranges and its slope and compass aspect; i.e., slopes facing the sun are drier and warmer. The physical structure of the underlying geological formations, plus historical events such as volcanic ash deposits, determine the type of soils present. Together, these factors have a decisive influence on the kinds of vegetation growing on a particular site and on the relative vigor of the plants, i.e., their ability to defend themselves against insects. Finally, the plant community and physical setting have important influences on the abundance of carnivorous organisms and pathogens, as well as on species that compete with or are allied to the pest herbivore. All these factors are included in the basic environment of the organism, and it is this basic environment which sets the stage on which the action of population change takes place (Fig. 4.1).

The basic environment may also be subjected to unpredictable variations in weather or other physical *disturbances,* such as droughts, extreme cold, fires, and volcanic eruptions. These disturbances may drastically affect the favorability of an organism's environment from time to time; e.g., the explosion of Mount St. Helens in Washington State covered large areas with abrasive volcanic ash and caused significant increases in insect death rates (i.e., dramatically reduced the favorability of the environment for many insects).

4.1.3. The Effects of Population Density

The favorability of an organism's environment may also be influenced by the density of the population (effect c in Fig. 4.1). For instance, an increase in population density will invariably reduce the amount of food and space available to each individual, even if the total quantities remain unaffected; e.g., less food will be available per individual when 100 larvae share 10 leaves than when 10 larvae share the same amount of food. Population density can, therefore, affect the favorability of the environment, or the "standard of living" for the individual organism, by reducing the availability of food, shelter, and hiding places, or by making the individual more susceptible to predators, parasites, and pathogens. As a result, individuals may die of starvation or predator attacks; even if they survive, the malnourished organisms will usually produce fewer offspring. The crowded conditions may also induce some individuals to leave in search of "greener pastures." As a result, the birth and immigration rates of the average individual will decline while the death and emigration rates rise, causing the density of the population to decrease (effects a and b, in Fig. 4.1). What we have, in effect, is a *negative feedback* loop in which the density of the population negatively affects the favorability of its environment, which then reduces the individual reproduction, survival, and/or immigration rates or increases the emigration rate; this feeds back to reduce the density of the population (loop a, b, c in Fig. 4.1). We will call this a negative ($-$) density-dependent feedback loop.

As we learned earlier (Chapter 3), negative feedback loops usually have a stabilizing influence. Thus, we would expect populations affected by ($-$) density-dependent feedback to reach an equilibrium position eventually where births and immigrations are balanced by deaths and emigrations (i.e., $b + i = d + e$). This basic notion is developed more formally in Fig. 4.2. To simplify the discussion, we assume that immigrations and emigrations are roughly equal and that population density has no effect on the individual birth rate. The individual death rate, however, is assumed to rise as the population becomes more dense because each organism will obtain less food and space (Fig. 4.2a). Equilibrium density E occurs when the birth and death rates are equal ($b = d$); below this density the population grows, because $b > d$, whereas above it the population

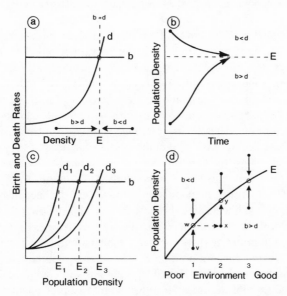

FIGURE 4.2. The effect of ($-$) density-dependent feedback, acting through the death rate d of individuals, on the growth of a hypothetical population when immigration equals emigration and the birth rate b is constant. (a) Equilibrium population density E where births = deaths ($b = d$) and below which populations grow ($b > d$), while above it they decline ($b < d$). (b) Dynamics of a population starting above or below equilibrium. (c) Effect of changing the basic properties of the environment from unfavorable (d_1 = higher death rate) to favorable (d_3 = lower death rate) on the equilibrium density. (d) Continuous relationship between environmental favorability and the equilibrium density showing dynamics after an environmental improvement x. See text for explanation.

declines ($b < d$). Thus, at whatever density the population starts, it will always converge on the equilibrium density E (Fig. 4.2b).

It is important to realize that although populations governed by ($-$) density-dependent feedback always move toward equilibrium, they do not necessarily remain at a constant level. For one thing, environments rarely remain constant over time. In addition, populations often overshoot or undershoot their equilibrium positions. Both phenomena tend to cause populations to oscillate to some degree around their equilibrium densities. Rather than expecting constant population densities, we should, therefore, expect populations to oscillate to varying degrees around their equilibrium levels but always to tend to return toward equilibrium when they are disturbed away from it.

Let us now try and visualize how changes in the basic environment may affect population equilibria. Obviously, the total quantity (or quality) of food and space, as well as the numbers of carnivores, competitors, and allies, can affect the individual death rate. For example, the death rate should rise more slowly in environments that contain greater quantities of food or food with better nutritional content or in which natural enemies and competitors are rare. As a result, the equilibrium point will occur at higher population densities (Fig. 4.2c). Thus, the basic environment has a decisive effect on the density of the equilibrium population.

Environmental conditions vary from place to place and from time to time. Variations in land form, for example, will create warmer and drier conditions on slopes that face the sun, and changing weather patterns will cause temporal variations in this condition. Equilibrium population densities will be expected to vary in space and time in response to these factors. An appreciation of the dynamic behavior of populations in changing environments can be obtained by plotting the equilibrium densities as a continuous function of basic environmental conditions (Fig. 4.2d). In this diagram the equilibrium densities obtained in Fig. 4.2c are plotted against the condition of the environment (poor–good) and the continuous relationship drawn by extrapolation. This *equilibrium function* separates the graph into zones of population growth and decline; in other words, populations above the line decline (because $b < d$) while below it they grow ($b > d$). For example, if we start with a population at v (in a poor environment), it will grow toward equilibrium at w. However, if the environment improves to x, the population will then converge on a new and higher equilibrium at y. In nature, where environments are rarely constant, populations will continuously be adjusting to new equilibrium densities. In other words, natural populations usually follow, or track, dynamic equilibria that change in response to basic environmental conditions.

4.1.4. Delayed Density-Dependent Feedback

It is also possible for population density to affect the basic properties of the environment (i.e., effect d in Fig. 4.1). For example, dense herbivore populations may cause the death of their food plants or may induce strong defensive responses in the plants being attacked; this will change the quantity and quality of the food supply (Chapter 3). In addition, natural enemy populations may increase when prey become abundant (Chapter 5). These density-dependent effects are different from those discussed previously because the characteristics of the basic environment are altered. For instance, when populations directly affect the favorability of their environments (effect c in Fig. 4.1), the consequences (e.g., lack of food, hiding places) are felt very quickly by individual organisms, and undernourished individuals will die rapidly (within a generation) from the effects of starvation or predators and diseases. When populations affect their basic environments, however, the consequences are often felt by future generations. For example, when the quantity or quality of food is reduced by intense feeding (e.g., trees are killed or defensive reactions stimulated), less total food may be available for succeeding generations. In addition, predators and parasites will often migrate into areas in which their prey are abundant, and the well-fed carnivores will also produce more offspring. As it takes a certain amount of time for natural enemies to find dense prey aggregations and to reproduce, the impact is also felt by future prey generations. Thus, the main difference between interactions d and c in Fig. 4.1 is that the former brings about changes in the abundance or quality of food and natural enemies, which then affect the reproduction and survival of future generations, while the latter only affects the birth and death rates of the current generation. This is an important distinction because we know that time delays in negative feedback loops often induce cyclic patterns of abundance. A more formal demonstration of the delayed feedback effect is presented in Fig. 4.3. The population starts at time zero below equilibrium and grows to its new density at time 1 (the solid arrow starting at time 0). In so doing, however, it reduces the favorability of the environment for the next generation (the dotted arrow). In this generation (time 1–2) the population only grows by a small number because it is very close to the equilibrium line E but, being even more dense, causes a further environmental deterioration (dotted arrow). Continuing with this line of reasoning, and noting that the environment improves when the population becomes sparse (time 4–6), we obtain a cyclic population trajectory (Fig. 4.3a,b). Thus, if the feedback loop e, a, b, d (Fig. 4.1) operates with a time delay, population density may cycle around its equilibrium level. In order to draw attention to this behavioral difference, we identify feedback loops that involve the basic environment as delayed negative ($-$) density-dependent.

4.1.5. Positive Density-Dependent Feedback

So far we have discovered two negative feedback loops regulating the dynamics of natural populations: (1) a rapid ($-$) density-dependent feedback loop operating directly through the favorability of the environment and maintaining a "tight" equilibrium, and (2) a delayed ($-$) density-dependent feedback loop operating indirectly through the basic environment and maintaining a

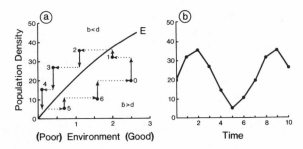

FIGURE 4.3. Population cycles may result when the favorability of the environment is affected by the density of preceeding insect generations. (a) Dynamics around the equilibrium line E: Starting at time zero in a favorable environment ($F = 2.5$), the population grows from 20 to 32 (solid arrow at 0), but in so doing causes a reduction in the favorability of the environment (from $F = 2.5$ to $F = 2$, dotted arrow) for the next generation (at time 1). The population then grows to 36 but in the process reduces environmental favorability even further ($F = 1.1$). In the third generation (time 2–3), the population decreases to 27, because it is above its equilibrium line, but is still dense enough to negatively affect the environment (to $F = 0.4$). This is followed by a further decrease (to 15) in the fourth generation (time interval 3–4) and a reduction in environmental favorability to $F = 0.1$. The population then declines to 5 in the fifth generation and is now so sparse that the environment improves to $F = 0.7$. This is followed by further improvements (to $F = 1.6$, then $F = 2.4$) as the population grows to its initial density of 20 at time 0. (b) The time trajectory of population density demonstrates the cyclical dynamics.

"loose" or cyclic equilibrium. There is, however, one more type of feedback that can occur when population density actually increases the favorability of an individual organism's environment. Consider, for example, a very sparse population in which the probability of an individual finding a mate and, consequently its birth rate, is very low (in this sense the environment is unfavorable because a resource, mates, is scarce). As the density of the population rises, contacts between individuals will increase as will the rate of mating and reproduction. In this instance, population density has a direct positive effect on the favorability of the environment (availability of mates), and this gives rise to a positive feedback loop; i.e., increased population density increases mating and reproduction, which then increases population growth rates. Positive feedback can lead to instability (see Chapter 3).

We can examine the effects of positive (+) density-dependent feedback more formally by constructing a simple model (Fig. 4.4). In this example we have assumed, for simplicity, that the individual death rate does not change with population density and that emigration and immigration are equal. The individual birth rate, however, is assumed to rise with population density up to a maximum value, the potential fecundity of the species (Fig. 4.4a). Once again we find an equilibrium population density, T, where births are equal to deaths. In this case, however, the equilibrium point is unstable because populations below this point decline continuously ($b < d$), while those above it grow continuously ($b > d$).

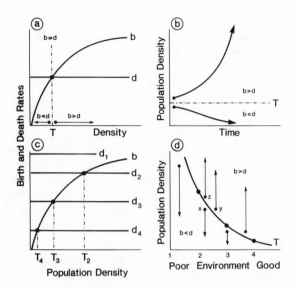

FIGURE 4.4. The effect of ($+$) density-dependent feedback on population dynamics. In this example the death rate d is constant, immigrations exactly balance emigrations, and the birth rate b increases with population density up to a maximum. (a) An unstable equilibrium density occurs at T, where $b = d$, because the population grows when it is above T ($b > d$) and declines when it is below T ($b < d$). (b) Divergent population dynamics resulting when the initial density is just above or just below the unstable threshold T. (c) Effect of different death rates, reflecting changing basic environmental conditions, on the location of the unstable threshold (d_1 = unfavorable to d_4 = favorable basic environments). (d) The continuous threshold function as it changes with basic environmental conditions showing how a small change in the environment from x to y, or an increase in the insect population from x to z, can result in radically different population dynamics.

This divergent dynamic behavior is illustrated in Fig. 4.4b. We see that populations that were initially very close to each other, but separated by equilibrium point T, diverge from each other as time progresses (compare with the convergent dynamics illustrated in Fig. 4.2b).

Because the unstable equilibrium point caused by ($+$) density-dependent feedback separates discrete zones of qualitatively different population behavior, it is sometimes called a *population threshold*. This draws attention to the fact that once populations are moved beyond the unstable threshold, say by human actions, they exhibit radically different patterns of change. In our example (Fig. 4.4a,b), the population automatically declines to extinction if it is reduced, say by harvesting, below the unstable threshold density T. For this reason, low-density unstable equilibria are often called *extinction thresholds*.

Extinction thresholds are extremely important to managers of natural populations, and especially to those concerned about endangered species. Passenger pigeons, which had complex colonial mating habits and hence rather high extinction thresholds, probably became extinct when hunting pressure reduced their populations below this critical threshold density. Other useful animals, such as whales, may be in danger of a similar fate. Insects, on the other hand, often have effective mate-finding pheromones and high mobility; because of this, they are less likely to be driven to extinction by human actions (i.e., they have very low extinction thresholds). Some insects such as aphids and scales even reproduce parthenogenetically (asexually); in these cases reproduction is possible from a single female.

The basic environment has an important influence on the position of the unstable threshold. In Fig. 4.4c, for example, we assume that the constant death rate is higher in an unfavorable environment (d_1) than in a favorable one (d_4). As we can see, the unstable threshold occurs at a low population density when the environment is favorable (T_4) and then rises as the basic environmental conditions deteriorate $(T_3$ and $T_2)$. In extremely unfavorable environments (d_1), moreover, the species always declines to extinction irrespective of its initial density because the number of deaths always exceeds births. These extremely poor environments determine the boundaries of species distributions in space.

We can also plot the unstable population threshold as a continuous function of the basic environment (Fig. 4.4d). This *threshold function* has important implications for population managers, as it specifies the environmental conditions under which populations will grow to very high densities (outbreak behavior) from those in which they remain at low densities or go locally extinct (endemic behavior). It also illustrates why such populations are very sensitive to environmental disturbances and insect migrations. For instance, if the population is at the position x in Fig. 4.4d, it will decline because it is below the unstable threshold line (i.e., $b < d$). However, if the environment improves, even by a relatively small amount (say x to y), the population will grow continuously. Alternatively, if a few insects were to migrate into the area, raising the density from x to z, the population would again move across the threshold into the zone of continuous growth. Thus, we can imagine how outbreaks starting in one locality, an *epicenter,* can spread by migrating to adjacent regions. Such outbreaks have been termed *eruptive,* drawing attention to their explosive and expansive nature; their unstable equilibria are termed *outbreak thresholds.*

Because (+) density-dependent feedback can have such a dramatic effect on the dynamics of insect populations, it is important to realize that other processes, besides mating, can be involved in these (+) feedback loops. For example, some bark beetle species are able to overwhelm more resistant hosts if their populations become large, thereby increasing the number of trees that can be successfully invaded. In effect, the food available per beetle, hence the favorability of the environment, is improved by increases in population density, creating a (+) density-dependent feedback loop.

Positive feedback can also occur when the probability of an individual being killed by predators or parasites decreases with population density. This may happen if the species forms defensive aggregations (flocks, schools, or swarms) or if predators become satiated or confused by dense prey populations (this topic is discussed in more detail in Chapter 5). In general, we should expect (+) feedback and the possibility of unstable population thresholds whenever there is an "advantage in numbers"—that is, when organisms cooperate with each other in offense or defense, or when high population densities provide individual organisms with an advantage in gaining food, escaping from predators or parasites, or in reproducing.

4.1.6. Density-Independent Factors

In the preceding sections we dealt almost exclusively with environmental factors that are affected in some way by population density. Some environmental variables are never influenced by the density of a particular population, whereas others may be affected at certain times but not at others. Environment factors that are not affected by population density are termed *density independent.*

The most obvious density-independent variables are components of the physical environment, such as site characteristics (e.g., elevation, aspect, slope) and weather (e.g., wind, temperature, precipitation). Biotic factors, however, may also act in a density-independent manner or may be affected by density under certain conditions but not under others. For instance, the food supply for most bark beetles is determined by external factors (e.g., disturbances such as lightning strikes, diseases, and wind storms) when the beetle population is sparse (Fig. 4.5a). Under these conditions, the food supply acts as a density-independent variable. However, when populations of "aggressive" bark beetle species become dense, the number of susceptible trees may increase as a function of population size because mass attack by large numbers of beetles enables them to overcome more resistant hosts (Fig. 4.5b; see also Table 2.1). At this time, the food supply is therefore positively affected by population density giving rise to an unstable (+) feedback loop. But as beetle populations continue to grow, more and more trees will be killed, and the food supply for future beetle generations will be reduced. This will give rise to a delayed (−) feedback loop (Fig. 4.5c).

Although density-independent variables are not involved in the feedback structure that regulates insect population density, they do play an important role in determining the level at which regulation occurs. For instance, the variable death rates shown in Fig. 4.2c and 4.4c would probably be caused by changes in density-independent variables. In these cases, it is the density-independent factors that determine the equilibrium densities, or set the level at which (−) density-dependent feedback regulates the population (Fig. 4.2c). They would also determine the threshold at which (+) density-dependent feedback initiates population explosions or collapses (Fig. 4.4c). In addition, unpredictable variations in density-independent factors can sometimes precipitate insect outbreaks or collapses when they disturb populations to one side or another of an unstable threshold (Fig. 4.4d).

a. Sparse beetle population

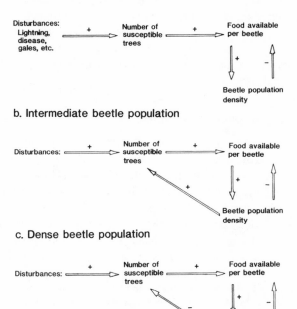

b. Intermediate beetle population

c. Dense beetle population

FIGURE 4.5. The number of trees susceptible to colonization by "aggressive" bark beetles can change from a density-independent variable (a) to a (+) density-dependent variable (b), to a delayed (−) density-dependent variable (c), depending on the density of the beetle population. See text for explanation.

4.1.7. Summary of Population Principles

The basic principles of population dynamics are crucial to understanding and managing forest insect pests. They are condensed and summarized below:

1. Population dynamics are entirely determined by the rates of birth, death, and movement into or out of the area in which the population exists. Populations grow when $b + i > d + e$, decline when $b + i < d + e$, and remain in equilibrium when $b + i = d + e$.
2. Individual birth, death, and movement rates are directly related to the favorability of the environment or to the availability of food, shelter, hiding, and nesting places, as well as the chances of escaping or avoiding predators, parasites, diseases, and competitors.
3. Environmental favorability can be negatively affected by population density because resources must be shared among more individuals as

populations become dense. This (−) density-dependent feedback leads to a decline in birth rates, an increase in death rates as population density rises, and often gives rise to convergent equilibrium population behavior. Equilibria are usually "tightly" regulated because the feedback occurs rapidly, usually within a generation.

4. Population density can also affect the basic properties of the environment in a negative way, i.e., when food is removed faster than it can be replaced or when natural enemies reproduce or immigrate into an area containing a dense prey population. But these effects are usually transmitted with a time delay and are therefore identified as delayed (−) density-dependent feedback. Delayed (−) feedback loops produce more "loosely" regulated equilibria and often give rise to population cycles.

5. The favorability of an organism's environment is sometimes related to population density in a positive way. This may happen when organisms cooperate with each other for mating, defense, or attack or when there are other advantages in numbers. Positive density-dependent feedback can create unstable population thresholds (extinction or outbreak thresholds) that separate distinctly different patterns of divergent population behavior (population collapses vs. explosions).

6. Density-independent variables, which do not respond to population density, are often involved in determining the density at which populations attain equilibria and at which unstable thresholds occur. They also act as disturbances that can precipitate outbreaks or population collapses.

4.2. PATTERNS OF DYNAMIC POPULATION BEHAVIOR

Although at first sight forest insect populations seem to exhibit an astounding array of dynamic behavior, our general analysis suggests that only three basic patterns should be observed: (1) relatively stable populations that remain close to equilibrium or that do not fluctuate violently around their equilibrium positions; (2) populations that exhibit violent cycles of abundance; and (3) populations that occasionally erupt and spread over large areas. In addition, we should expect variations in basic environmental conditions to affect the level of population equilibrium and the amplitude of population cycles and to precipitate eruptions. Any particular insect population may exhibit one, two, or all three types of behavior but will often be characterized by a single dominant pattern. In the next section, we will examine the dynamics of natural forest insect populations and will attempt to interpret their behavior in the light of our general analysis.

4.2.1. Relatively Stable Populations

Of the many herbivorous insects inhabiting forest environments, most seem to remain at very low densities where damage to the resource rarely or never occurs. In Table 4.1, for example, we see that 74 insect species were collected from *Eucalyptus* stands in Australia, but only four species have ever been recorded in outbreak numbers. The total number of insects was also low, averaging around 20 per kilogram of host foliage, and these only ate about 2% of the annual leaf production. Populations of these herbivorous insects must therefore be regulated at very low densities by (−) density-dependent feedback. Consequently forests remain green most of the time, even though they are inhabited by numerous species of herbivorous insects—the species are many, but their populations are usually sparse.

Unfortunately, direct long-term data on the fluctuations of sparse forest insect populations are rather rare. This is because most research has been done on outbreak species at times when outbreaks were in progress. The few reported studies indicate that relatively stable populations oscillate to varying degrees around their apparent equilibrium densities (Fig. 4.6). Some populations seem to exhibit cyclic tendencies (Fig. 4.6a and 4.6b top), suggesting the presence of delayed (−) density-dependent feedback, whereas others oscillate more sharply (Fig. 4.6b bottom and Fig. 4.6c), indicating the action of rapidly responding (−) feedback. The mountain pine beetle data (Fig. 4.6c) are interesting because they also illustrate how the equilibrium population density can change quite suddenly from time to time. Apparently the basic density-independent environment changed in 1959 and the beetle population was forced to a much lower equilibrium level in subsequent years.

4.2.2. Violent Population Cycles

Although cyclical tendencies are often observed in some relatively stable populations (Fig. 4.6a), other forest insect populations go through extremely

TABLE 4.1
Herbivorous Insects Collected in the Crowns of *Eucalyptus* Stands with Data on Leaf Production and Consumption[a]

Insect orders	Number of species[b]	Density/kg foliage	Leaf consumption (all species) (kg/ha per yr)	Leaf production (kg/ha per yr)
Defoliators				
Lepidoptera	35 (1)	5.5		
Coleoptera	31 (1)	6.3	40.6	1903.3
Phasmatodea	1 (1)	<0.1		
Sucking insects				
Hemiptera	7 (1)	8.0		
Total	74 (4)	19.9	Leaf consumption = 2%	

[a]From Ohmart *et al.* (1983a,b).
[b]Number of outbreak species in parentheses (Ohmart, personal communication).

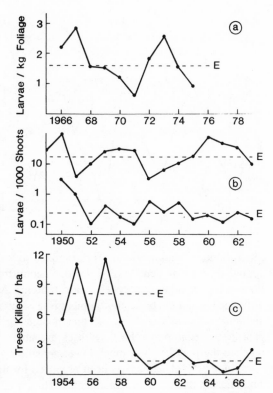

FIGURE 4.6. Dynamics of some relatively stable forest insect populations. (a) Larch budmoth caterpillars, *Zeiraphera diniana*, at Lenzburg, Switzerland. (From Auer, 1975.) (b) Two species of caterpillars, *Bupalus piniarius* (top) and *Thera firmata* (bottom), feeding on Scots pine in Holland. (After Klomp, 1968.) (c) Lodgepole pines killed by the mountain pine beetle, *Dendroctonus ponderosae*, on Starvation Ridge, Glacier National Park. (From Tunnock, 1970.)

violent cycles of abundance (Fig. 4.7). The remarkable features of these cycles are their regularity and their incredible amplitude. For example, peak population densities of the larch budmoth in the Swiss Alps occur every 9 or 10 years, and these peak densities are often as much as 30,000 times greater than the minimum densities (Fig. 4.7).

Regular population cycles are usually caused by delayed ($-$) density-dependent feedback. Most cyclical forest insects are leaf or needle feeders that severely defoliate their host trees but rarely kill them. However, the quantity and quality of host foliage (the basic environment) is usually altered by heavy feeding, and this can give rise to delayed ($-$) feedback. In the case of the larch budmoth, for example, the needles produced by larch trees in the years after heavy defoliation are shorter and more fibrous, contain lower nitrogen concentrations, and are covered by a film of resin. These changes in the nutritional quality of the foliage reduce the survival of budmoth larvae and the fecundity of surviving adult females in the years after heavy defoliation because it takes several years for the needles to return to their more nutritious condition.

It is interesting that although violent population cycles are observed in certain regions or environments they do not occur in others. At Lenzburg,

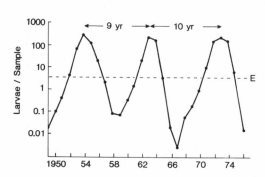

FIGURE 4.7. Population dynamics of the larch budmoth in the Engadine Valley of Switzerland. Mean numbers of budmoth larvae per kilogram of larch foliage, shown on a logarithmic scale because of the enormous differences between the troughs and peaks of the cycles (up to 30,000-fold increases and decreases). E = the mean density of larvae per kilogram over the whole period (E ~2.4). (Redrawn from Fischlin and Baltensweiler, 1979; see also Baltensweiler et al., 1977, for a review of the population ecology of the larch budmoth.)

Switzerland (elevation 500 m), for example, budmoth populations exhibit a mere sixfold variation in density (Fig. 4.6a), as compared with the 30,000-fold variations in the Engadine Valley (elevation 1850 m). A plausible explanation for this difference is that the basic environment is less suitable for budmoth reproduction and survival near Lenzburg, so that population growth rates are lower, and parasites, predators, or host defenses are able to respond rapidly enough to regulate the population at much lower densities.

Another explanation has been proposed for differences in the amplitude of autumnal moth cycles in Scandinavia. This moth periodically defoliates birch stands growing in northern Scandinavia and in mountainous regions of Norway but remains at relatively stable densities in the warmer areas. These differences in population behavior can be explained by the observation that birch leaves respond to defoliation very quickly (no delay) in the low-altitude and southerly stands, whereas defensive responses are delayed until the following year in the colder regions. Cycles may result in the colder regions, therefore, because of time delays in the negative feedback between host plant and insect populations.

Because local environmental conditions play such a critical role in determining the amplitude of population cycles, outbreaks of cyclical insects usually reoccur in the same places and rarely spread into adjacent regions. As a general rule, high-amplitude cycles can be expected in those environments that are more favorable for insect reproduction and/or survival or that are less favorable for the host plant or natural enemies.

It is also interesting, and somewhat amazing, that cycles of different and widely separated populations often cycle in synchrony with each other. For example, outbreaks of the Douglas-fir tussock moth in western North America seem to be synchronized with each other in almost all areas, with the possible exception of Arizona and New Mexico (Fig. 4.8). Tussock moths, like the larch bud moths and autumnal moths, cycle with a periodicity of 9–10 years. It is even

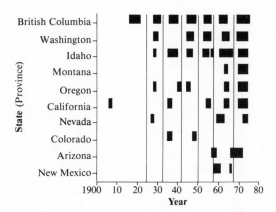

FIGURE 4.8. Years that large-scale outbreaks of the Douglas fir tussock moth, *Orgyia pseudotsugata*, occurred in western North America. (From Clendenen, 1975, as reported in Brookes *et al.*, 1978; USDA Forest Service drawing.)

more amazing, although perhaps coincidental, that cycles of these three widely separated species are also approximately synchronized. The cause of cycle synchronization is unknown but it seems likely that weather disturbances are involved. For example, an occasional severe winter may reduce populations simultaneously over a wide area so that they all begin cycling in unison. If synchrony really occurs between very widely spaced populations, global or cosmic forces, such as sun-spot activity, may be involved in cycle synchronization.

4.2.3. Population Eruptions

Some insect populations remain at very low densities for long periods of time but then suddenly explode to extremely high densities and spread over large forested areas. This kind of behavior is demonstrated, for example, by the mountain pine beetle population in Glacier National Park. Before 1970 the beetle population was restricted to a relatively small area (162 hectares), and the total number of trees killed each year varied from less than 1 to 11 per hectare (Fig. 4.6c). After 1972, however, the beetle population expanded rapidly until, by 1977, almost 58,000 hectares were infested and more than 10 million trees were killed in 1977 alone (Fig. 4.9). This incredible population eruption represents a 5 millionfold increase in mountain pine beetle killed trees over a 10-year period (1967–1977).

Erupting populations often increase in size logarithimically or exponentially. Thus, the average growth rate of the population can be estimated by plotting the logarithm of population size against time (see the insert in Fig. 4.9). The slope of this line (r in Fig. 4.9) provides an estimate of the exponential growth rate of the population and is related to the individual rates of birth, death, and movement in the following way:

$$r = \log_e(b - d + i - e + 1) \tag{4.3}$$

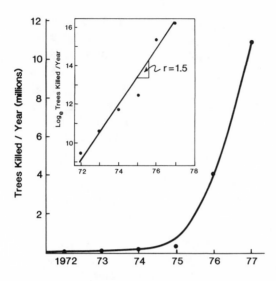

FIGURE 4.9. Growth of a mountain pine beetle population expressed as the number of lodgepole pines killed by the beetle per year in Glacier National Park from 1972 to 1977 shown on arithmetic and natural logarithmic scales. (Data from McGregor *et al.*, 1982.)

Eruptive outbreaks frequently start in local environments, or *epicenters,* that are very favorable for the reproduction and survival of the pest, but then spread into less favorable environments as the emigrating insects exceed the outbreak thresholds of adjacent stands. This characteristic expansion over large forested areas is illustrated in Fig. 4.10. In this case, spruce budworm epicenters seem to occur in mixed conifer–hardwood forests that have been subjected to ecological disturbances such as selective logging or fire, but then spread north and west into pure spruce–fir stands.

From our theoretical discussion we know that population eruptions are associated with (+) density-dependent feedback. We also know that there must be an unstable equilibrium point separating the relatively stable low-density dynamics shown in Fig. 4.6c from the explosive outbreak of Fig. 4.9 (see, e.g., Fig. 4.4d). In the case of the mountain pine beetle, this (+) feedback seems to be initiated when beetle populations are able to overcome vigorous trees by rapid mass attack (Fig. 4.5b; see also Table 2.1). In this way, the number of trees available for attack increases as population density rises. If this hypothesis is correct, the beetle population must have been too sparse before 1967 to infest the more resistant pines and was therefore restricted to a few trees weakened by lightning strikes, root pathogens, and so forth (Fig. 4.5a). Sometime after this date, however, the population must have surpassed the outbreak threshold, and this allowed it to spread rapidly throughout the pine stands of the Park.

In the case of the spruce budworm, on the other hand, the positive feedback seems to result when dense budworm populations saturate the feeding responses of insectivorous birds. In other words, the risk of an individual budworm being eaten by birds decreases with density because the predators become satiated more

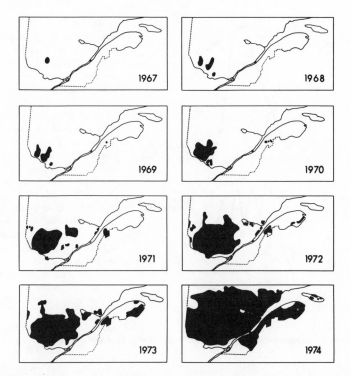

FIGURE 4.10. The eruptive spread of the spruce budworm, *Choristoneura fumiferana*, from epicenters in southern Quebec, Canada, to eventually cover most of the spruce–fir forest of the province. (Reproduced from Hardy *et al.*, 1983, with permission of the Society of American Foresters.)

quickly. This "advantage in numbers" can give rise to (+) density-dependent feedback and unstable outbreak thresholds (see Chapter 5 for a more detailed discussion of "escape" from natural enemies).

Eruptive outbreaks of the mountain pine beetle and spruce budworm can have a devastating effect on their host plants, often killing more than ninety percent of the mature stems in a given stand. Because the food supply for these insects is almost completely depleted in a few generations, the population must go through a pulselike cycle in any one locality (Fig. 4.11a). The cycle is, in fact, caused by delayed (−) density-dependent feedback *via* rapid environmental deterioration. Other eruptive species may have less serious effects on their hosts; in these cases, outbreaks may persist for a number of years in a given area (Fig. 4.11b). For example, eruptions of defoliators, such as the gypsy moth, that feed on deciduous trees may continue in one stand for a number of years because leaves are replaced each year and trees only die after many years of heavy defoliation.

Figure 4.11 also illustrates two of the important characteristics of eruptive

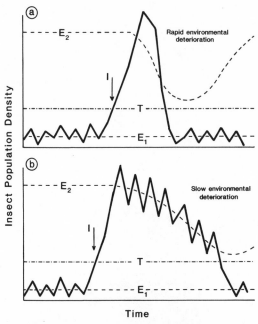

FIGURE 4.11. Dynamics of population eruptions in a given place. Both populations are regulated at low densities (E_1) by negative density-dependent feedback with natural enemies or food supplies provided that their densities do not exceed the unstable threshold T. When an immigration of insects I raises the local populations above their outbreak thresholds T, they escape from their enemies or overwhelm resistant hosts and expand under the influence of ($+$) density-dependent feedback. After the populations reach epidemic levels, however, population (a) rapidly depletes its food supply and/or natural enemy populations quickly increase and this delayed ($-$) density-dependent feedback forces the pest population back to its endemic equilibrium. Population (b), on the other hand, causes less severe and more gradual changes to the favorability of its environment and so declines more slowly in a series of oscillations.

populations. First, that outbreaks can be initiated in a given stand by insect migrations. It is this property that enables pest eruptions to spread in successive wavelike pulses from their epicenters (Fig. 4.10). Second, we can see how outbreaks can be initiated at their epicenters by temporary environmental disturbances that either lower the outbreak threshold or increase the resident insect population (e.g., Fig. 4.4d).

4.3. CLASSIFICATION OF FOREST PEST OUTBREAKS

From our analysis of forest insect population dynamics emerges an elementary scheme for classifying pest outbreaks according to their tendency to spread from epicenters or to arise and subside in place, and according to the relative stability of their equilibria. For example, some insect populations are regulated in a relatively stable condition by fast acting ($-$) density-dependent feedback. But their equilibrium levels may be affected by changes in density-independent environmental variables. In such cases, changes in the basic environment lead, automatically, to corresponding changes in the equilibrium density of the insect population (Fig. 4.12a). Because the average densities of these populations reflect local environmental conditions, or respond in a graded manner to local

FIGURE 4.12. Types of forest insect outbreaks. (a) Sustained pest gradients occur in environments that remain consistently favorable (right), or pulse gradients are induced when environments in particular areas change from low to high favorability and back (left to right and vice-versa). (b) Cyclical gradients occur when favorable environments amplify population cycles caused by delayed ($-$) density-dependent feedback. (c) Eruptive outbreaks are induced at irregular intervals in moderately favorable environments by temporary environmental disturbances that elevate population densities above their unstable thresholds, or are continuous in highly favorable environments. Eruptive outbreaks invariably spread from epicenters and may be further subdivided into pulse eruptions (c_1), sustained eruptions (c_2), cyclical eruptions (c_3), and permanent eruptions (c_4).

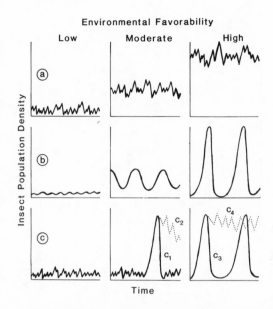

environmental changes, and because outbreaks do not spread into less favorable environments, we term such population changes *gradients* as distinct from the eruptive, spreading type.

Insect environments may vary in both time and space. For example, pine plantations on relatively dry sites (e.g., slopes facing the sun) provide more favorable environments for shoot moths; consequently, shoot damage tends to be consistently higher on these sites (Table 4.2a). In stands growing on moister sites, however, shoot moths rarely attain damaging numbers. When populations remain consistently high in certain favorable environments we will call them *sustained gradient outbreaks* (Fig. 4.12a, right column).

Fir engraver beetle populations also exhibit gradients, being dependent on the abundance of severely weakened fir trees; these "non-aggressive" bark beetles are unable to overcome the resistance of their hosts because they are repelled by even small quantities of defensive chemicals (see Table 2.1). The number of severely weakened trees, however, is strongly affected by time-varying phenomena such as root decay, defoliation, and droughts. With this beetle, we see gradient outbreaks that persist at one locality for a short time but then subside to low densities as the environment returns to normal. We will call these short-termed population increases in response to temporary environmental changes *pulse gradient outbreaks* (Table 4.2b).

TABLE 4.2
Examples of Types of Forest Insect Outbreaks

Insect species	Feeding habitat	Effects on host plants	Effectiveness of enemies	Causes of outbreaks	Outbreak class	Reference
a. Western pine-shoot borer *Eucosma sonomana*	In terminals	Reduced height growth	Ineffective	Dry sites	Sustained gradient	Stoszek (1973)
b. Fir engraver beetle *Scolytus ventralis*	Under bark	Kills weak trees	Ineffective	Drought, defoliation, root pathogens	Pulse gradient	Berryman (1973)
c. Red pine cone beetle *Conophthorus resinosae*	In cones	Reduced reproduction	Ineffective	Large cone crops	Pulse gradient	Mattson (1980)
d. Ambrosia beetle *Trypodendron lineatum*	In wood; on fungus	None; infests dead trees	Ineffective	Logging, windthrow	Pulse gradient	Prebble and Graham (1957)
e. Larch budmoth *Zeiraphera diniana*	On foliage	Weakens trees and re-duces growth	Effective at high densities	High elevation sites	Cyclical gradient	Baltensweiler *et al.* (1977)
f. Douglas-fir tussock moth *Orgyia pseudotsugata*	On foliage	Weakens trees, re-duces growth, some mortality	Effective at high densities	Dry sites, stand structure	Cyclical gradient	Mason and Luck (1978)
g. Autumnal moth *Epirrita autumnata*	On foliage	Weakens trees, re-duces growth, some mortality	Effective at high density	Northerly latitude and high elevation	Cyclical gradient	Haukioja (1980)
h. Mountain pine beetle *Dendroctonus ponderosae*	Under bark	Kills trees	Ineffective	Droughts, beetle migration, stand age and diameter	Pulse eruption	Berryman (1982)
i. Eastern spruce budworm *Choristoneura fumiferana*	On foliage	Kills many trees	Effective at low density	Warm weather, moth migrations	Pulse eruption	Clark *et al.* (1979)
j. Gypsy moth *Lymantria dispar*	On foliage	Weakens trees, re-duces growth, some mortality	Effective at low density	Moth migration, dry sites, human inter-ference	Sustained eruption	Campbell and Sloan (1977, 1978)
k. European woodwasp *Sirex noctilio*	In wood	Kills trees in Austral-asia	Effective at low density	Spreading pathogen, woodwasp migra-tion	Pulse eruption	Taylor (1983)
l. Jack pine sawfly *Neodiprion swainei*	On foliage	Weakens trees, re-duces growth, some mortality	Effective at low density	Dry weather, sawfly migration	Sustained eruption	McLeod (1979)

Both sustained and pulse gradient outbreaks are usually associated with insects that live inside, rather than on, their host plant, and that do not normally cause host mortality or, if they do kill their host, only attack severely weakened plants. Because of their cryptic habits, these insects are not normally subjected to heavy parasitism, predation, or disease, and their population levels are determined, to a large extent, by the abundance of food. These pests are typically cone and seed feeders, tip and shoot borers, wood borers, and bark beetles that can only infest dead or dying twigs, branches, and stems (Table 4.2a–d).

Insect populations that affect the quantity or quality of their basic environments and, thereby, induce delayed (−) feedback, sometimes exhibit regular periodic outbreaks in highly favorable environments (Fig. 4.12b). Because the amplitude of the cycles also reflects a graded response to environmental conditions, these outbreaks are called *cyclical gradients*.

Cyclical gradient outbreaks are identified by their regularity, usually every 8–11 years (Fig. 4.7) and by their tendency to be synchronized over relatively large areas (Fig. 4.8) and to occur, time and again, in the same geographic regions or on the same sites. In addition, cyclical gradients do not spread far from their points of origin, are usually associated with defoliating insects, particularly moths and butterflies (Lepidoptera) that rarely cause extensive mortality to their host plants, and the outbreaks are often terminated by natural enemies or host defensive responses (Table 4.2e–g).

Some forest insect populations remain at relatively stable levels for long periods of time but then suddenly erupt to very high densities (Fig. 4.12c). These pest eruptions usually begin in particular localities (epicenters), but then spread over large forested areas (Fig. 4.10). Eruptions are often triggered at their epicenter by sudden disturbances or gradual changes in the basic environment and then spread as insects emigrate into surrounding regions.

Insect populations that undergo eruptive outbreaks are characterized by unstable outbreak thresholds that separate low-density populations from explosive outbreaks. The threshold is created by (+) density-dependent feedback, often involving cooperative attack and defense behavior. For example, "aggressive" bark beetles are able to overwhelm resistant hosts cooperatively; some woodwasps inoculate pathogens or toxins into healthy trees and thereby increase the availability of weak trees in which they can breed; some sawflies cooperatively defend themselves against predators; and some caterpillars have a greater chance of surviving predation by birds and/or small mammals as their populations become more dense (Table 4.2h–l).

Within the group of eruptive pests we can identify four outbreak subclasses, depending on the severity of their impact on their host plant populations and the effectiveness of their natural enemies. Those outbreaks that cause the death of a large percentage of their host plant populations or that are rapidly suppressed by natural enemies or host defensive responses often exhibit *pulse eruptions* in any one locality (Fig. 4.12c$_1$). On the other hand, outbreaks that do not severely impact the plant population or that only kill trees after many years of heavy attack, may persist for a long time before subsiding to their low-density equilibrium. We call these *sustained eruptions* (Fig. $4.12c_2$). Finally, *cyclical eruptions* may occur in highly favorable environments if the outbreak has little affect on plant survival and is terminated by host defensive responses or natural enemies (Fig. $4.12c_3$), whereas *permanent eruptions* are theoretically possible if natural enemies or host defenses are ineffective, and host survival is not impacted at high population densities (Fig. $4.12c_4$). However, because eruptive outbreaks that are not terminated by other factors should eventually have an impact on host plant survival, permanent eruptions do not seem to be likely in nature.

TABLE 4.3
Key to the Outbreak Classes

Class	Outbreak characteristics	Outbreak class	
1a	Spread out from local epicenters to cover large areas of forest and usually last for several to many insect generations.	*Eruptions*	2
1b	Do not spread far from their points of origin and are associated with particular sites or major disturbances, but the latter always subside when the environment returns to normal.	*Gradients*	4
2a	Go through a single pulselike cycle at any one place, often being terminated by food depletion, host-defensive reactions, or natural enemies.		3
2b	Persist at high densities for several to many years at any one location, and host plants only die after many years of attack if at all (Fig. $4.12c_4$).	*Sustained eruption*	
3a	Occur at regular intervals, often 8–11 years apart, and never cause severe or widespread mortality to the host plant population (Fig. $4.12c_3$).	*Cyclical eruption*	
3b	Occur at irregular intervals and often cause severe and rapid mortality to the host plant population or are quickly terminated by natural enemies (Fig. $4.12c_1$).	*Pulse eruption*	
4a	Occur at irregular intervals following major environmental disturbances or are permanently associated with particular site and/or stand conditions.		5
4b	Occur at regular intervals, usually every 8–11 years, rarely cause extensive morality to the host plant population, are often associated with particular site and/or stand conditions, and are usually terminated by host-defensive responses or natural enemies (Fig. 4.12b).	*Cyclical gradient*	
5a	Occur more or less continuously on particular sites and/or stands (Fig. 4.12a).	*Sustained gradient*	
5b	Occur at irregular intervals, following major environmental disturbances or outbreaks of other organisms and subside soon after the environment returns to normal.	*Pulse gradient*	

We have now identified six types of insect outbreak, divided into two broad classes. The eruptive outbreaks are characterized by their self-perpetuating spreading nature, while gradient outbreaks arise and subside in place in response to external environmental conditions. The outbreak subclasses reflect the tendencies of populations to persist in place for several insect generations, to decline rapidly from outbreak levels, or to cycle with regular periodicity. The key shown in Table 4.3 classifies specific pest outbreaks according to these characteristics.

Before leaving this discussion, we need to consider the problem that arises when insects are accidentally introduced into a foreign environment. This new environment may be highly favorable for the alien insect because the plant hosts often lack effective defenses and predators, parasites, or pathogens are often absent. Under these conditions, permanent outbreaks are likely to occur (i.e., the right-hand column of Fig. 4.12). Although exotic pests create a particular kind of problem, their population dynamics are frequently eruptive at first because they tend to spread rapidly from a point of origin, or epicenter. After a period of time, however, they may subside and settle into a relatively stable pattern or may exhibit one of the three basic outbreak types. For example, populations of the larch casebearer, winter moth, and several other pests imported into North America seem to have settled into relatively stable patterns after initial periods of population growth and spread and after the importation of natural enemies from their native home (see Chapter 8).

We should also note that the primary aim of forest pest management should be to create conditions where all insect populations are in a relatively stable low-density state (Fig. 4.12, left-hand column). This objective is only possible if the manager understands the basic patterns of pest population dynamics and the feedback processes that give rise to these patterns. In addition, the forest manager who is able to identify the class of outbreak exhibited by a particular pest will be prepared to manipulate the causal environmental variables in an attempt to keep the pest population in a nonoutbreak condition (to the left of Fig. 4.12). These applications are discussed in Part III.

REFERENCES AND SELECTED READINGS

Auer, C., 1975, Jährliche und langfristige Dicteveränderungen bei Lärchen-wicklerpopulationen (*Zeiraphera diniana* Gn.) ausserhalb des Optimumgebietes, *Bull. Soc. Entomol. Suisse* **48**:47–58. (population cycles of the larch budmoth)

Baltensweiler, W., Benz, G., Bovey, P., and Delucchi, V., 1977, Dynamics of larch bud moth populations, *Annu. Rev. Entomol.* **22**:79–100. (population cycles)

Benz, G., 1974, Negative Rückkippelung durch Raum-und Nahrungskonkurrenz sowie zyklische Veränderung der Nahrungsgrundlage als Regelprinzip in der Populations-dynamik des Grauen Lärchenwicklers, *Zeiraphera diniana* (Guenée), *Z. Angew. Entomol.* **76**:196–228. (delayed negative feedback in larch budmoth system)

Berryman, A. A., 1973, Population dynamics of the fir engraver, *Scolytus ventralis* (Coleoptera: Scolytidae). I. Analysis of population behavior and survival from 1964 to 1971, *Can. Entomol.* **105**:1465–1488. (gradient outbreaks)

Berryman, A. A., 1981, *Population Systems: A General Introduction,* Plenum Press, New York. (Chapters 1, 2, and 3: population concepts and dynamics; Chapter 5: spatial dynamics and eruptions)

Berryman, A. A., 1982, Mountain pine beetle outbreaks in Rocky Mountain lodgepole pine forests, *J. For.* **80:**410–413. (eruptive population dynamics)

Berryman, A. A., 1983, Population dynamics of bark beetles, in: *Bark beetles in North American Conifers. A System for the Study of Evolutionary Biology* (J. B. Mitton and K. B. Sturgeon, eds.), pp. 264–314, University of Texas Press, Austin. (bark beetle eruptions)

Berryman, A. A., 1986, The theory and classification of outbreaks, in: *Insect Outbreaks* (P. Barbosa and J. C. Schultz, eds.), Academic Press, New York (in press).

Berryman, A. A., and Baltensweiler, W., 1981, Population dynamics of forest insects and the management of future forests, *Proc. XVII IUFRO World Congr. Kyoto* **2:**423–430. (review of population research)

Brookes, M. H., Stark, R. W., and Campbell, R. W., 1978, The Douglas-fir tussock moth: A synthesis, U. S. Forest Service Technical Bulletin 1585, 331 pp. (population cycles)

Campbell, R. W., 1974, The gypsy moth and its natural enemies, U. S. Forest Service, Agriculture Information Bulletin 381, 27 pp. (exercises)

Campbell, R. W., and Sloan, R. J., 1977, Release of gypsy moth populations from innocuous levels, *Environ. Entomol.* **6:**323–330. (exercises)

Campbell, R. W., and Sloan, R. J., 1978, Numerical bimodality in North American gypsy moth populations, *Environ. Entomol.* **7:**641–646. (exercises)

Campbell, R. W., Miller, M. G., Duda, E. J., Biazak, C. E., and Sloan, R. J., 1976, Man's activities and subsequent gypsy moth egg-mass density along the forest edge, *Environ. Entomol.* **5:**273–276. (exercises)

Clark, W. C., Jones, D. D., and Holling, C. S., 1979, Lessons for ecological policy design: A case study of ecosystem management, *Ecol. Mod.* **7:**1–54. (spruce budworm eruptions).

Fischlin, A., and Baltensweiler, W., 1979, Systems analysis of the larch bud moth system. Part 1: the larch–larch bud moth relationship, *Bull. Soc. Entomol. Suisse* **53:**273–289. (role of host tree in population cycles)

Graham, K., 1963, *Concepts of Forest Entomology,* Reinhold, New York. (Chapter 6: population concepts)

Hardy, Y. J., Lafond, A., and Hamel, L., 1983, The epidemiology of the current spruce budworm outbreak in Quebec, *For. Sci.* **29:**715–725.

Haukioja, E., 1980, On the role of plant defenses in the fluctuation of herbivore populations, *Oikos* **35:**202–213. (role of host tree in population cycles)

Klomp, H., 1968, A seventeen year study of the abundance of the pine looper *Bupalus piniarius* L. (Lep. Geometridae), in: *Insect Abundance* (T. R. E. Southwood, ed.), pp. 98–108, Blackwell, Oxford. (population equilibria)

Larsson, S., and Tenow, O., 1980, Needle-eating insects and grazing dynamics in a mature Scots pine forest in Central Sweden, *Ecol. Bull. (Stockh.)* **32:**269–306. (sparse stable populations and needle consumption)

Mason, R. R., and Luck, R. F., 1978, Population growth and regulation, in: The Douglas-fir tussock moth: A synthesis, U. S. Forest Service Technical Bulletin 1585. (Brookes, M. H., Stark, R. W., and Campbell, R. W., eds.), pp. 41–47.

Mattson, W. J., 1980, Cone resources and the ecology of the red pine cone beetle, *Conophthorus resinosae* (Coleoptera: Scolytidae); *Ann. Entomol. Soc. Am.* **73:**390–396. (cone beetle exercise)

McGregor, M. D., Oakes, R. D., and Meyer, H. E., 1982, Status of Mountain Pine Beetle, Northern Region, 1982, U. S. Forest Service, Northern Region, Report No. 83-16. (bark beetle eruption)

McLeod, J. M., 1979, Discontinuous stability in a sawfly life system and its relevance to pest management strategies, in: Selected Papers in Forest Entomology from XV International Congress of Entomology, (W. E. Waters, ed.), pp. 71–84, U. S. Forest Service, Gen. Tech. Rept., WO-8.

Morris, R. F., 1959, Single-factor analysis in population dynamics, *Ecology* **40**:580–588. (exercise)

Ohmart, C. P., Stewart, L. G., and Thomas, J. R., 1983a, Leaf consumption by insects in three *Eucalyptus* forest types in southeastern Australia and their role in short-term nutrient cycling, *Oecologia (Berl.)* **59**:322–330. (sparse, stable populations)

Ohmart, C. P., Stewart, L. G., and Thomas, J. R., 1983b, Phytophagous insect communities in the canopies of three *Eucalyptus* forest types in south-eastern Australia, *Aust. J. Ecol.* **8**:395–403.(sparse stable populations)

Prebble, M. L., and Graham, K., 1957, Studies of attack by ambrosia beetles in softwood logs on Vancouver Island, British Columbia, *For. Sci.* **3**:90–112.

Raffa, K. F., and Berryman, A. A., 1983, The role of host plant resistance in the colonization behavior and ecology of bark beetles (Coleoptera: Scolytidae), *Ecol. Monog.* **53**:27–49. (cooperative behavior that leads to eruptive outbreaks)

Schwerdtfeger, F., 1935, Studien über den Massenwechsel einiger Forstschädlinge, *Z. Forst. Jagdw.* **67**:15–540. (population equilibria)

Schwerdtfeger, F., 1941, Über die Ursachen des Massensechsels der Insecten, *Z. Angew. Entomol.* **28**:254–303. (population equilibria)

Shepherd, R. F., 1977, A classification of western Canadian defoliating forest insects by outbreak spread characteristics and habitat restrictions, in: *Insect Ecology* (H. M. Kulman and H. C. Chiang, eds.), pp. 80–88, Technical Bulletin 310, Agricultural Experiment Station, University of Minnesota. (an outbreak classification system)

Stoszek, K. J., 1973, Damage to ponderosa pine plantations by the western pine-shoot borer, *J. For.* **71**:701–705.

Taylor, K. L., 1983, The *Sirex* woodwasp: Ecology and control of an introduced forest insect, in: *The Ecology of Pests: Some Australian Case Histories.* (R. L. Kitching and R. E. Jones, eds.), pp. 231–248, CSIRO, Melbourne.

Tunnock, S., 1970, A chronic infestation of mountain pine beetles in lodgepole pine in Glacier National Park, Montana, *J. Entomol. Soc. BC* **67**:23. (equilibrium dynamics)

Varley, G. C., Gradwell, G. R., and Hassell, M. P., 1973, *Insect Ecology: An Analytical Approach*, Blackwell Scientific Publications, London. (Chapter 8: examples of forest insect population dynamics)

CHAPTER 5

NATURAL ENEMIES OF FOREST INSECTS

Herbivorous forest insects are preyed upon by a variety of insectivorous organisms, including birds, mammals, reptiles, spiders, mites, other insects, nematodes, fungi, bacteria, and viruses. These carnivorous organisms can play an important role in maintaining sparse pest populations or in suppressing incipient outbreaks. The forest manager should therefore understand how these organisms operate in forest ecosystems.

In Chapter 3 we noticed that interactions with carnivores can produce stabilizing negative feedback loops (see Fig. 3.5). For such loops to be present, however, certain conditions must be met in the interaction between carnivore and prey: (1) the natural enemy must have a negative impact on the prey population by causing the death or lowering the reproduction of attacked prey; and (2) the number of prey attacked must increase in direct relationship to the density of the prey population. The first condition is usually met because most parasites, predators, and pathogens cause the death of their insect prey. The second condition, however, may or may not be met and depends on the characteristics of the natural enemy responses to prey density. For this reason, we need to examine the density-dependent responses of natural enemies in some detail.

5.1. NATURAL ENEMY RESPONSES TO PREY DENSITY

The responses of natural enemies to the density of their prey can be separated into two types. In the first type, called the *functional response*, each individual natural enemy responds by attacking more prey as the density of its prey increases (Fig. 5.1). In the second, or *numerical response*, the number of natural enemies increases as prey population density rises. This increase in numbers may be attributable to a *reproductive response*, wherein better fed

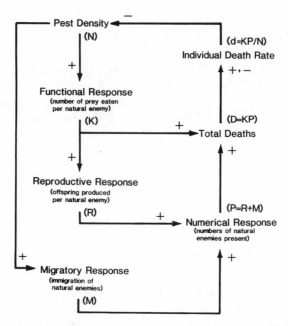

FIGURE 5.1. Components of the response of natural enemies to the density of their prey. As prey density increases more are attacked by each natural enemy (functional response) and this may increase the prey's death rate (first feedback loop; $N \rightarrow K \rightarrow D \rightarrow d \rightarrow N$). Greater nutrition of natural enemies through the functional response gives rise to more offspring (reproductive response), and natural enemies move into regions of high prey density (migratory response). The overall increase in the numbers of natural enemies (numerical response) causes a reduction of prey density through the functional response (second feedback loop; $N \rightarrow (R + M) \rightarrow P \rightarrow D \rightarrow d \rightarrow N$).

natural enemies produce more offspring, or a *migratory response*, wherein they move into areas of high prey density.

If we examine Fig. 5.1 closely, we see that there are two potential feedback loops in the interaction between natural enemy and prey populations, either one of which could be capable of regulating the density of an insect prey. The first feedback loop involves the functional response alone, while the second involves both functional and numerical responses. An important question is therefore whether the functional response can itself regulate pest populations in the absence of a change in natural enemy numbers. This question is important because some natural enemies, especially vertebrates, are limited by nesting places or territorial behavior and so do not have strong numerical responses to the abundance of their food.

5.1.1. Functional Responses

The general functional response is characterized by an increasing rate of attack as the density of the prey population rises, but this eventually reaches a plateau as the natural enemy becomes satiated (Fig. 5.2). There are three variants of the functional response but we will only be concerned with the two forms usually associated with predators and parasites of forest insects. In the first type, the rate of attack approaches its maximum at an ever-decreasing rate, or it decelerates continuously as prey density increases (Fig. 5.2c). We will call this a *cyrtoid functional response* to refer to its convex shape. The deceleration of the attack rate results because the natural enemy spends more time in subduing, eating, and digesting its prey as more are caught so that less time is available for search and attack. This kind of response is often associated with arthropod predators and parasites.

The second kind of functional response is called *sigmoid* to refer to its S-shaped form (Fig. 5.2s). In this type, the attack rate *accelerates* with increasing prey density at first but then slows down as the natural enemy becomes satiated. Sigmoid functional responses are commonly found in vertebrate predators that form *searching images* for certain prey species, particularly those that are "tasty" and more abundant. These predators tend to switch their attention to the more abundant prey species, and it is this switching that causes the functional response to accelerate at first. As the prey become more dense, however, the response decelerates in the same manner as the cyrtoid type because of the time constraints on search and attack activities.

Having described the two basic kinds of functional responses, we now need to investigate their role in pest population regulation. In order to simplify the argument, let us assume that the individual prey birth rate b is constant and that all deaths are caused by a single natural enemy. Thus, the number of prey killed per unit of time K is determined by the natural enemy's functional response. The individual death rate can then be calculated as the number killed divided by the density of the population (i.e., $d = K/N$). When this death rate is calculated for a pest population attacked by a natural enemy with a cyrtoid functional response, we notice that the individual death rate declines continuously with prey density and that the equilibrum point T is always unstable (Fig. 5.3a).

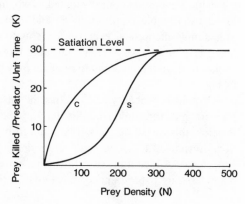

FIGURE 5.2. Two types of functional response of natural enemies to the density of their prey populations: c, cyrtoid response, where the attack rate rises at an ever-decreasing rate with prey density; s, sigmoid response, where the attack rate accelerates at first but then decelerates at high prey densities.

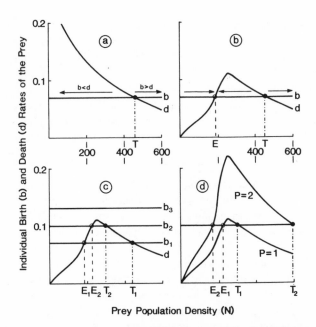

FIGURE 5.3. Equilibrium conditions imposed on a prey population by different natural enemy functional responses when the individual prey birth rate is assumed constant. Calculations of individual death rates are made directly from Fig. 5.2 by assuming that $d = K/N$. (a) Cyrtoid functional responses produce an unstable threshold T. (b) Sigmoid functional responses can produce stable E and unstable T equilibrium points. (c) Increasing the constant individual birth rate of the prey ($b_1 \rightarrow b_3$) leads to escape from regulation by the natural enemy. (d) Increasing the number of foraging natural enemies P under the assumption that the individual prey death rate is $d = PK/N$ reduces the equilibrium level E and increases the escape threshold T.

Figure 5.3b shows the individual death rate curve caused by a natural enemy with a sigmoid functional response. In this case the individual death rate increases at first in direct relationship to population density but then declines as the natural enemy becomes satiated. Two equilibrium points can be formed, a low-density stable equilibrium E, and a high-density unstable equilibrium T. This simple theoretical exercise leads us to conclude that natural enemy functional responses always create unstable thresholds and that sigmoid functional responses have the potential to regulate prey populations at very low densities (E in Fig. 5.3b).

You have probably realized that the unstable prey equilibrium T is in fact a prey outbreak threshold (see Chapter 4). When the density of the prey population is below this threshold, it will be reduced by the action of natural enemy functional responses to a low equilibrium density (Fig. 5.3b) or to local extinction (Fig. 5.3a). Above this critical density, however, the prey population will grow unhindered by natural enemy functional responses.

Although the functional response of some natural enemies has the potential to regulate prey populations at low densities, certain conditions have to be met before regulation is possible. For example, we can see that prey populations can escape regulation if their birth rates are too high (Fig. 5.3c) or if the number of natural enemies is too low (Fig. 5.3d). It is also evident that, even if equilibrium conditions exist, the likelihood of prey outbreaks increases when natural enemy populations are reduced or when pest birth rates rise because the distance between the unstable and stable equilibrium is reduced (Fig. 5.3d). (Note that the probability of an outbreak occurring is higher when $P = 1$ because smaller environmental disturbances are sufficient to displace the population from E_1 to T_1.)

The environment in which natural enemy and prey populations coexist obviously plays a crucial role in determining whether regulation is possible. Changes in the environment that are unfavorable for the natural enemy can result in a decrease in the size of its population and the escape of the prey from low-density regulation (Fig. 5.3d). Conversely, a change that favors the prey can cause an increase in its birth rate or a decrease in deaths from other causes, and this may have a similar result (Fig. 5.3c). The lesson for the forest manager is obvious: Forestry practices that improve the environment for the pest, or reduce natural enemy populations, can directly precipitate pest outbreaks or increase the risk of outbreaks following natural disturbances (i.e., lower the outbreak threshold).

5.1.2. Numerical Responses

Numerical responses occur when natural enemy populations increase, either through reproduction or migration, as a result of increases in the density of their prey. Reproductive numerical responses may be expected whenever natural enemies are limited by their food supply, the prey. In these cases an increase in the abundance of the prey results in more food for foraging predators and a corresponding increase in their survival and reproduction. Other natural enemies may be limited by the availability of water, nesting sites, territories, or other resources and will not normally increase in numbers as a result of higher prey densities. They may, however, exhibit migratory numerical responses at certain times. For example, natural enemies may move away from watering places, nesting sites, or even territories at certain times of the day, or in certain seasons, to concentrate in places where their prey is abundant. In addition, mobile animals tend to move around more when their prey is scarce but less when their prey is abundant, and these trivial movements tend to create concentrations of natural enemies in regions of high prey density.

Numerical responses vary greatly from one group of natural enemies to another, but their importance in regulating prey populations is determined, to a large extent, by the rate at which the response occurs relative to the population growth rate of the prey. Thus, prey populations are more likely to be regulated at low densities by natural enemy numerical responses if

1. Each natural enemy is capable of producing many offspring (high fecundity or birth rate b), so that their populations can change rapidly in response to changes in prey density (fast reproductive numerical responses are partly dependent on the individual birth rate b).
2. Natural enemies have short life cycles, relative to their prey, so that their populations can respond more rapidly to changes in prey density (i.e., numerical responses are faster if the natural enemy produces several generations for each prey generation).
3. Natural enemies have alternative foods, or can enter dormancy when their food is scarce, so that their populations do not get very low when a particular prey is scarce (reproductive responses are more rapid when populations are large because the rate of change is proportional to bP, where P is the size of the natural enemy population and b is the birth rate).
4. Natural enemies are not limited by territories, nesting places, or resources other than food, at least not until their populations become very large.
5. Natural enemies have well developed sensory mechanisms for searching out and finding their prey and are highly mobile, so that they can quickly locate dense prey aggregations (fast migratory numerical response).
6. Prey do not have very high birth or survival rates in the absence of predation (i.e., the prey's environment is not too favorable).

It is important to realize that numerical responses differ from functional responses in the speed of their reaction. When the density of prey increases, hungry natural enemies immediately eat more. As such, the sigmoid functional response is an immediate reaction to changes in prey density and should maintain a "tight" prey equilibrium (see Chapter 4). On the other hand, it takes a certain amount of time for increased food intake to be transformed into more offspring (Fig. 5.1) and for natural enemies to locate and aggregate at locations where their food is abundant. The numerical response, therefore, usually lags behind changes in prey population density, and the length of this delay will be related to the breeding interval, mobility, and searching capabilities of the natural enemy species. As we found earlier, delays in the reaction of density-dependent negative feedback often result in cyclical population dynamics (Chapters 3 and 4). Thus, we should not be surprised to see regular cycles of abundance in those pest species that are regulated by the numerical responses of their natural enemies (Fig. 5.4).

With this general introduction to the subject of pest population regulation by interactions with their natural enemies, we shall now turn our attention to the specific characteristics of various natural enemy groups in an attempt to understand their roles in regulating populations of forest insects.

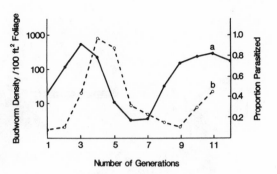

FIGURE 5.4. Population cycles of the black-headed budworm (a) in the spruce—fir forests of New Brunswick apparently caused by the numerical responses of parasitic insects (b). (After Morris, 1959; see also Chapter 6, Fig. 6.7.)

5.2. VERTEBRATE PREDATORS

Numerous forest-dwelling vertebrates are exclusively insectivorous, or at least use insects as part of their diet (Fig. 5.5). Of these, birds are the most diverse and abundant group. Woodpeckers are adapted for feeding on insects living under bark or burrowing in wood; i.e., bark beetles, flatheaded and round-headed borers, carpenter ants, and so on. Other birds, such as tits, nuthatches, cuckoos, juncos, chickadees, warblers, grosbeaks, sparrows, and flycatchers, feed on the exposed larvae of defoliating insects and on the adult stages of almost all forest insects. Mammals are the second most important group, usually feeding on insect larvae, pupae, or adults while they are present on or in the ground. Of the mammals, shrews and mice are probably the most significant in reducing forest insect populations. However, skunks, racoons, opossums, and bats may also take their share. Reptiles such as frogs, snakes, and lizards also have insectivorous diets, but their specific habitat requirements rarely bring them into contact with most of the important forest pests.

Vertebrate predators have a number of characteristics that influence their particular role in regulating forest pest populations. First, they are general predators that normally feed on a large number of different insect pests, as well as beneficial species. Second, they have voracious appetites (Table 5.1). For example, individual birds may consume up to 100 insects a day, giving them the potential to kill more than 30,000 insects per year. Shrews and mice may be even more ravenous, some consuming as many as 300 insects per day, or up to a maximum of 100,000 per year (see Table 5.1). Third, most vertebrate predators have sigmoid functional responses that result from their complex learning abilities. Intelligent vertebrate predators learn to attack, or form a "searching image" of, those palatable prey species that are most frequently encountered. They also learn which specific habitat harbors the more abundant species. This leads them to concentrate their attacks on the more abundant insect species in their diet. It is this switching of feeding activity to the more dense prey populations that causes the functional response to accelerate over the lower prey density ranges

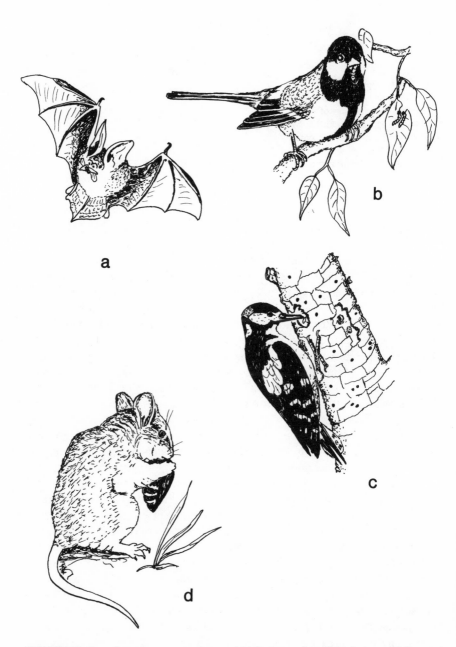

FIGURE 5.5. Some insectivorous vertebrates. (a) The bat, a free-flight forager; (b) the great tit, a foliage forager; (c) the woodpecker, a trunk forager; and (d) the mouse, a ground forager.

TABLE 5.1

Characteristics of Some Insectivorous Organisms

Insectivore	Prey	Attack potential per predator	Fecundity /female per year	Functional response	Numerical response	Source
Birds						
Hairy woodpecker	Bark beetle larvae	95/day	4	Sigmoid?	Weak?	Otvos 1965
Olivebacked thrush	Budworm pupae	74/day	9	Sigmoid?	Weak?	Mook and Marshall 1965
Robin	Sawfly larvae	6000/year	8	Sigmoid	Weak?	Buckner 1966
Evening grosbeak	Budworm larvae and pupae	20000/year	5	Sigmoid?	Migratory	Takekawa and Garton 1984
Mammals						
Shorttailed shrew	Sawfly pupae	320/day 9000/year	16	Sigmoid	None	Holling 1959 Buckner 1966
Masked shrew	Sawfly pupae	114/day 5000/year	15	Sigmoid	Sigmoid	Holling 1959 Buckner 1966
White-footed mouse	Gypsy moth pupae	28/day	12	Sigmoid?	Weak?	Smith and Lautenschlager 1981
Insects						
Black-bellied clerid	Bark beetle larvae and adults	170/year	1000	Cyrtoid	Strong?	Berryman 1966, 1967
Ichneumonid wasp	Budworm larvae	<140/year	140	Cyrtoid	Strong?	Miller 1960
Tachinid fly	Winter moth larvae	<1300/year	1300	Sigmoid	Strong?	Embree 1966

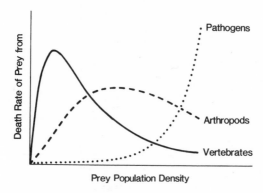

FIGURE 5.6. The relative effect of different natural enemy groups on the death rate of gypsy moth larvae at different densities. (Modified from Campbell, 1975.)

(Fig. 5.2s) and to endow them with the potential to regulate prey populations at very low densities (Fig. 5.3b).

On the negative side, vertebrate predators do not normally have rapid reproductive numerical responses. Many species are limited by factors others than the abundance of food, such as nesting places or territories. In these cases, predator numbers are often independent of the density of their prey. Even in those species that are food limited, the reproductive responses are often ineffective because the predators are limited by alternative foods at times of the year when insect prey are unavailable, or because their response is too slow to control the explosive growth of insect populations; i.e., the maximum birth rates of vertebrates are much lower than those of insects (Table 5.1).

Some vertebrate predators, however, have good migratory numerical responses. Birds, for example, often concentrate in large numbers in areas where their insect prey are most abundant. These migratory responses may help prevent pest populations from reaching outbreak densities but are probably ineffective in altering the course of outbreaks. At such times so many insects are present that even the concentrated feeding of migratory birds has little impact on population growth.

In general, vertebrate predators are most effective at limiting the growth of insect pest populations when the pests are relatively sparse (Fig. 5.6). Because of their voracious appetites and sigmoid functional responses, they may be capable of regulating pest populations at extremely low densities. For this reason they deserve more attention from the forest manager than is usually given them. Management procedures that improve the effectiveness of vertebrate predators, mainly by increasing their numbers, will be considered in Chapter 8.

5.3. ARTHROPOD PREDATORS

Populations of herbivorous forest insects are attacked by large numbers of predatory arthropods, particularly spiders (Arachnida) and mites (Acarina) of the subphyllum Chelicerata, and insects belonging to the orders Coleoptera (beetles), Diptera (flies), Hymenoptera (wasps and ants), Hemiptera (true bugs), Orthoptera (preying mantids), Neuroptera (lace-wings), and Mecoptera (scorpion flies) (Fig. 5.7). Each forest insect pest is preyed upon by as many as forty different predatory species, and these may have a profound influence on the survival rate of the pest. But in order for the predators to be able to regulate pest populations at low densities, they must respond functionally and/or numerically to the density of their prey. In those predators which have been intensively studied, the functional response is usually of the cyrtoid type (Table 5.1). In

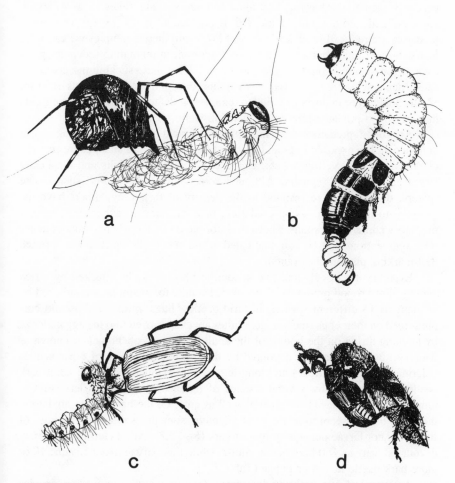

FIGURE 5.7. Some predaceous arthropods. (a) Spider feeding on a gypsy moth larva. (b) Ostomatid beetle larva feeding on a bark beetle. (c) *Calosoma* beetle feeding on a gypsy moth larva. (d) Clerid beetle feeding on an adult bark beetle. (Not drawn to scale; a and c after Campbell, 1975.)

addition, the attack potential of arthropod predators is much lower than that of vertebrates (Table 5.1). For these reasons, we would not expect arthropod predators to be very effective at regulating the density of their prey through the action of the functional response alone. On the other hand, arthropod predators usually have high reproductive potentials and are often highly mobile, leading us to suspect that they may be capable of regulating pest populations through their numerical responses.

The best examples of control of pest populations by the action of arthropod predators are to be found in agriculture. For example, an exotic scale insect in the citrus groves of Southern California was controlled by the introduction of a lady beetle (Coccinellidae) from Australia, and sap-sucking mites in Washington apple orchards are normally controlled by predaceous mites (except when these predators are reduced by insecticide sprays). From these examples we can infer that arthropod predators are particularly important in regulating sedentary pests such as sap-sucking aphids, scales, and mites. That these sucking insects rarely cause damage to forests, except when introduced from other lands (balsam woolly adelgid) or in dusty or polluted regions (predators are disturbed), suggests that arthropod predators are doing their job well.

Arthropod predators may also play important roles in regulating insect defoliators. The exposed eggs and larvae of these insects are attacked by an array of predatory ants, yellowjackets, and beetles, while the adults and dispersing larvae are caught by spiders. Although some of these predators, such as the spiders, do not seem to respond to the density of their prey, others have the potential to do so. The social predators, in particular, are expected to exhibit migratory numerical responses because information on the whereabouts of dense prey aggregations can be communicated at the nest (yellowjackets) or through trail-marking pheromones (ants).

Even cryptic insects that live inside their hosts may be attacked by a rich variety of arthropod predators. Some bark beetles, for example, are attacked by as many as 15 different species; histerid beetles, bugs, mites, and pseudoscorpions feed on their eggs (the wingless mites are even carried into the egg galleries by hitching a ride on the bodies of the adult beetles—a phenomenon known as phoresy); beetle larvae are devoured by the predaceous larvae of other beetles (Cleridae and Ostomatidae) and long-legged files (Dolichopodidae); adult bark beetles are attacked by several species of predaceous beetles (Cleridae, Ostomatidae, Cucujidae, Histeridae) as well as ants. Of the bark beetle predators, the clerid beetles seem to be the most effective, sometimes reaching densities of forty predator larvae per square foot of bark (Fig. 5.8). Adult clerid beetles may devour as many as 160 bark beetle adults, while their larvae may consume 10 or more bark beetle larvae or pupae (Table 5.1).

As a general rule, arthropod predators feed on many different prey species (i.e., they are polyphagous), have cyrtoid functional responses, and have relatively small appetites. They may, however, have strong migratory or reproduc-

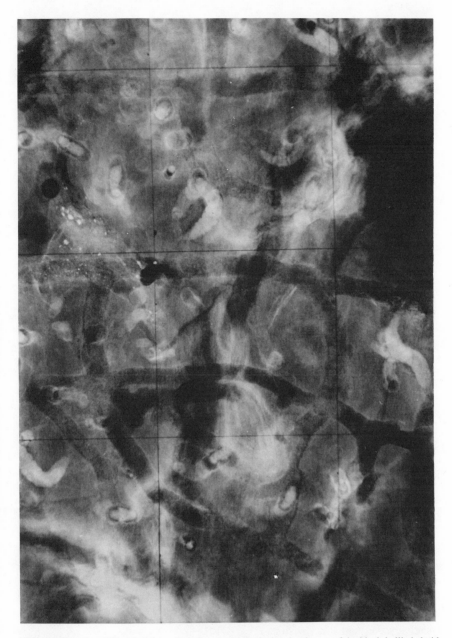

FIGURE 5.8. X-ray film of a piece of ponderosa pine bark showing larvae of the black-bellied clerid (large elongate objects) foraging for western pine beetle larvae, pupae, and adults (smaller objects). Note empty tunnels where bark beetles have been eaten. Clerid predators are most effective against bark beetles, such as the western and southern pine beetles, which live in the dry outer bark of the tree, often attaining densities of 40 or more clerid larvae per square foot. There is also evidence that clerids respond numerically to the density of these bark beetles. (See Berryman *et al.,* 1970 and Moore, 1972.). By permission of the Entomological Society of Canada.

tive numerical responses. For these reasons, we might expect them to be most important in regulating pest populations at intermediate densities (Fig. 5.6) and, in some cases, to cause prey population cycles.

5.4. ARTHROPOD PARASITES

In contrast to predators, arthropod parasites live for most of their lives on or within their insect prey. Thus, an individual parasite usually consumes no more than a single prey during its life span. In some of the smaller species, many parasites of the same species may develop within or on a single host (super-parasitism), or two or more parasites of different species may occasionally develop in one host individual (multiple parasitism).

Most of the parasites of forest insect pests belong to the insectan orders Hymenoptera (wasps) and Diptera (flies) (Fig. 5.9). Because they have rather unique life cycles and habits, which always lead to the death of their host, insect parasites of other insects are usually called *parasitoids*. Female parasitoids are well adapted for locating and parasitizing their prey, being highly mobile, equipped with well developed sensory organs, and some having well-developed ovipositors with which to pierce their host or the bark and wood in which their hosts live (Hymenoptera: Fig. 5.9b,c). Other parasitoids lay their eggs on the outside of their prey (Fig. 5.9a,d) or distribute them on the prey's food supply where they may be eaten by the feeding pest, or the newly hatched larvae may search out and penetrate into the host's body (some Diptera).

Because of their unique way of life, the functional and numerical responses of parasitoids are practically indistinguishable. Obviously, each female parasitoid will be able to find and parasitize more hosts as the density of the host population increases (functional response), but this activity immediately gives rise to more parasites (numerical response). Thus, the two responses merge together so that the attack potential (satiation level) of the parasitoid is identical to its reproductive potential (fecundity) (Table 5.1). In most cases studied, the ovipositional response of parasitoids is of the cyrtoid type and, hence, it is probably the numerical response which is most important in pest population regulation. Because of this we might expect pest populations regulated by parasitoids to exhibit cyclic dynamics (Fig. 5.4). There is at least one example, however, of a tachinid fly parasitoid which has a sigmoid functional response (Table 5.1) and this may be why its host, the winter moth, is regulated at relatively constant low densities (Fig. 5.10).

Insect parasitoids have certain characteristics that sometimes enable them to regulate pest populations at very low densities. First, many parasitoids are adapted to attack a particular prey species (monophagous) or a relatively narrow range of prey species (oligophagous). Thus, their biology and behavior are often finely tuned to that of their prey. Second, they have extremely sensitive sensory adaptations for searching out the habitat of their prey, as well as the specific prey within

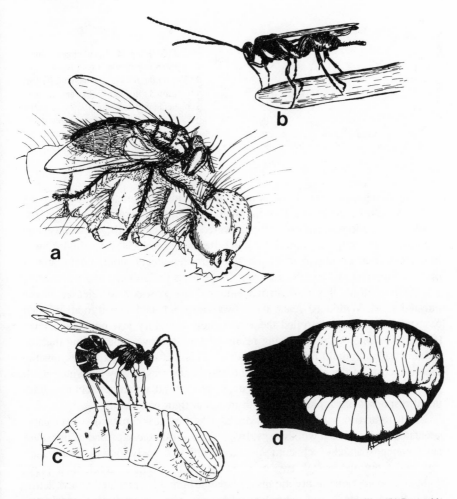

FIGURE 5.9. Some insect parasitoids. (a) Tachinid fly ovipositing on a tent caterpillar. (b) Braconid parasite about to oviposite in a caterpillar mining a larch needle. (c) Ichneumon wasp parasitizing a gypsy moth pupa. (Redrawn from Campbell, 1975.) (d) Braconid larva feeding on a bark beetle larva.

its habitat. Third, parasitoids often have high reproductive potentials (Table 5.1), and some adults kill additional hosts on which to feed before oviposition. Many parasitoids have short generation spans relative to their hosts and this, together with their high fecundity, enables them to respond quickly to changes in prey density (fast numerical response). Finally, parasitoids are not normally limited by resources other than their food supply, unless their numbers get very large, and this enables them to increase to high numbers and to suppress dense host populations.

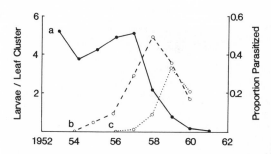

FIGURE 5.10. Regulation of winter moth populations (a) in Nova Scotia after the release of a tachinid fly parasitoid in 1954 (b) and a ichneumonid wasp in 1956 (c). (After Embree, 1966 who also showed that the tachinid had a sigmoid functional response.)

The best proof of the ability of insect parasitoids to regulate pest populations at low densities is to be found in the biological control literature. Although the problem of exotic pest control will be covered later (Chapter 8), we will use one example to illustrate the point (Fig. 5.10). Beginning in 1954, a tachinid fly was introduced from Europe in an attempt to control the winter moth, a defoliator of oaks that arrived in Nova Scotia sometime during the 1930s. Two years later, another parasitoid, this time an ichneumonid wasp, was also released in the infested areas. Within 10 years, these two parasitoids had spread to all infested oak stands and had controlled the winter moth at a very low population level (Fig. 5.10). One of the surprising results of this study was the finding that the tachinid had a sigmoid functional response. This fact may be partly responsible for the success of the control project and may be the reason why the winter moth population has remained at such low densities with no apparent cyclicity (functional responses are rapid and tend to maintain tight equilibria).

All herbivorous forest insects are attacked by a complex of insect parasitoids. Even the siricid woodwasps, living deep within the sapwood of trees, are not immune to attack—ichneumonid wasps with extremely long ovipositors are capable of boring through two or three inches of wood to reach their prey. Although populations of cryptic insects are not normally regulated at low densities by parasitoids, these natural enemies may play an important role in the eventual suppression of outbreaks. On the other hand, parasitoids probably play critical roles in the regulation of some defoliating caterpillars, sawflies, leaf beetles, and other insect pests exposed to their attack. Parasitoids are often influenced by conditions of their environment, and some species require alternative insect hosts or other food such as nectar or pollen. Diversity of environmental conditions usually increases the effectiveness of parasitoids with diverse habitat requirements.

As might be expected, some parasitoids have also evolved to attack those that feed on herbivorous insects. These secondary parasitoids, or hyperparasites, can potentially upset the balance between primary parasitoid and herbivorous host and thereby precipitate outbreaks of the pest.

5.5. PATHOGENS

Herbivorous forest insects suffer from a number of diseases resulting from infections by pathogenic microorganisms; i.e., fungi, bacteria, viruses, protozoa, and nematodes. Pathogenic infections may cause the death of their insect hosts or may cause debilitating diseases that reduce fecundity or make the insects more susceptible to other mortality factors.

Fungi, particularly those of the genera *Entomophthora* and *Beauveria,* are virulent and aggressive pathogens that usually cause the death of infected hosts. The fungi, however, usually die along with their host and, therefore, are not transmitted directly from one host generation to the next. Instead, airborne spores are released from fruiting bodies protruding from the dead insect and contact with a new host is purely a random process. Because of this inefficient transmission mechanism, and because fungal epidemics are critically dependent on rather precise conditions of temperature, humidity, and host density, they are rarely involved in the maintenance of low-density pest populations. On the other hand, fungal pathogens are often responsible for the collapse of agricultural pest outbreaks in humid regions and may sometimes play a similar role in forest ecosystems. Fungi may also be important in reducing the survival of insects that spend long periods of time in the soil, where conditions are favorable for fungal growth.

Bacteria have been found infecting almost all forest insects and some of them, like the well-known *Bacillus thuringiensis* (Fig. 5.11a), are virulent pathogens. Most bacteria, however, are not primary pathogens but rather cause disease symptoms or death of insects weakened by other stresses. The virulent pathogens, which cause the premature death of their hosts, are not transmitted from parent to offspring, hence have the same transmission problems encountered by pathogenic fungi. It seems unlikely that bacterial pathogens play a significant role in the regulation of low-density pest populations, and I am unaware of any examples of naturally occurring bacterial epizootics in forest insect populations. However, bacterial insecticide formulations have been used successfully to suppress outbreaks of defoliating caterpillars (see Chapter 9).

Forest insects are frequently infected by protozoan parasites. Microsporidia, in particular, have been found parasitizing both bark beetles and defoliators (Fig. 5.11b). Infection by these protozoa does not normally result in the death of the host but rather in reduced vigor, fecundity, and resistance to other stress agents. Protozoa are apparently transmitted from parent to offspring, hence may be quite prevalent even in sparse populations. Nevertheless, there is no evidence that these microorganisms are capable of regulating pest populations at low densities.

Viruses are one of the most common, and certainly the most devastating, natural pathogens of forest insect defoliators. Most insect viruses belong to the genus *Baculovirus* and are of the nuclear polyhedrosis or granulosis types

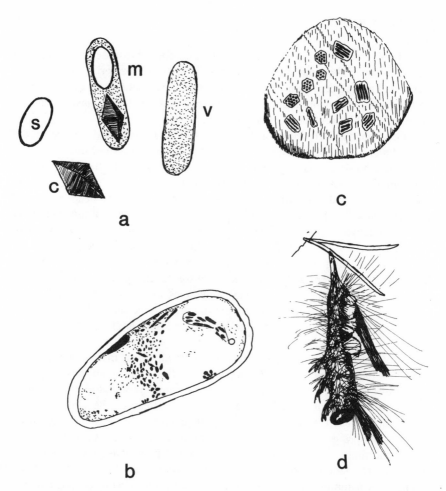

FIGURE 5.11. Some pathogenic microorganisms. (a) Cells of *Bacillus thuringiensis;* immature vegetative cell (v), mature cell with spore (m), protein crystal (c), and reproductive spore (s). (After Harper 1974.) (b) Spore of a microsporidian parasite of the pine weevil. (After Purrini, 1981.) (c) Cross section of Douglas-fir tussock moth virus polyhedron containing bundles of virus rods. (From Thompson and Hughes, 1977.) (d) Tussock moth larva killed by viral infection showing characteristic drooping, flacid posture. (From Thompson and Hughes, 1977.)

(Fig. 5.11c,d). Viruses are transmitted from individual to individual through contamination of the foliage and sometimes the eggs. Defoliator larvae killed by virus infections are flacid, their internal tissues liquified, and they typically hang upside-down from foliage or twigs (Fig. 5.11d). When these dead insects rupture, their contents contaminate foliage on which uninfected larvae are feeding. Eggs laid on bark, foliage or ground may also be contaminated in a similar manner.

Spread of viruses may also be aided by parasitoids or predators that become contaminated when attacking infected hosts and by the feces of birds that have fed on infected larvae.

The virulence of insect viruses seems to be affected by the density of the host population, probably because insects under nutritional stress are more susceptible, and also because transmission is more effective among crowded individuals (Fig. 5.6). Except for one case, in which an accidentally introduced virus was partly responsible for controlling the exotic European spruce sawfly, viruses

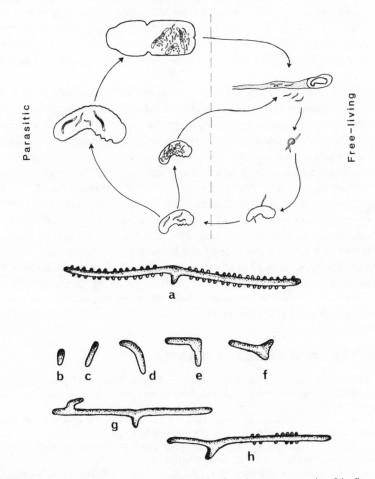

FIGURE 5.12. Life cycle of the nematode *Sulphuretylenchus elongatus*, a parasite of the fir engraver beetle, and egg galleries produced by uninfected (a) and nematode infected (b–h) females. Note abnormal galleries and reduced fecundity (lack of egg nitches) of nematode infected females. (After Ashraf and Berryman, 1970; by permission of the Entomological Society of Canada.)

do not seem to be a significant factor in the maintenance of forest pest populations at low-densities. Rather, their major role seems to be in suppressing outbreaks once the insect population has reached a relatively high density.

The nematodes, although not strictly pathogens, are common internal parasites of forest pests. Some nematode infections do not seem to have a significant impact on their host, while others can seriously affect vigor, fecundity, and survival (Fig. 5.12). Nematode parasites are not transmitted directly from parent to offspring but have free-living stages that search out and penetrate host individuals (Fig. 5.12). This may be why nematodes have successfully adapted to parasitize cryptic insects, such as bark beetles and woodwasps, whose offspring live in close proximity within infested trees. Nematodes may parasitize a large proportion of some bark beetle populations but their role in population regulation is not clear. In the case of the introduced European woodwasp in Australia, however, nematode parasites played an important role in regulating the pest (see Chapter 11).

Unlike the vertebrate and arthropod predators and parasitoids, most pathogenic microorganisms do not have highly developed sensory and locomotory adaptations for searching out and attacking their prey. Rather, they disperse passively in the environment (e.g., as spores) or are transmitted from parent to offspring by egg contamination, by phoresy, by contaminated parasitoids or predators, or random searching in a confined habitat. For this reason, we rarely expect pathogens to be effective at regulating their host populations at very low densities. On the other hand, pathogens have enormous reproductive potentials and very short generation spans, and many can remain for long periods of time in a dormant state (e.g., resistant spores). They can also spread rapidly once host populations reach high densities, and some become more virulent under crowded conditions. For these reasons, pathogens are very effective at suppressing pest populations once their densities become relatively high (Fig. 5.6) and, although they are unlikely to prevent outbreaks, they can be extremely powerful tools in suppressing outbreaks (Chapter 9).

REFERENCES AND SELECTED READINGS

Ashraf, M. and Berryman, A. A., 1970, Biology of *Sulphuretylenchus elongatus* (Nematoda, Sphaerulariidae), and its effects on its host, *Scolytus ventralis* (Coleoptera, Scolytidae), *Can. Entomol.* **102**:197–213.

Berryman, A. A., 1966, Studies on the behavior and development of *Enoclerus lecontei* (Wolcott), a predator of the western pine beetle, *Can. Entomol.* **98**:519–526.

Berryman, A. A., 1967, Estimation of *Dendroctonus brevicomis* (Coleoptera, Scolytidae) mortality caused by insect predators, *Can. Entomol.* **99**:1009–1014.

Berryman, A. A., Otvos, I. S., Dahlsten, D. L., and Stark, R. W., 1970, Interactions and effects of the insect parasite, insect predator, and avian predator complex, in: Studies on the Population Dynamics of the Western Pine Beetle, *Dendroctonous brevicomis* LeConte (R. W. Stark and D. L. Dahlsten, eds.), pp. 147–154, University of California Agricultural Service, Berkeley.

Buckner, C. H., 1966, The role of vertebrate predators in the biological control of forest insects, *Annu. Rev. Entomol.* **11**:449–470.

Campbell, R. W., 1975, The gypsy moth and its natural enemies, U. S. Department of Agriculture, Information Bulletin No. 381, 27 pp.

Campbell, R. W., Torgerson, T. R., and Srivastava, N., 1983, A suggested role for predaceous birds and ants in the population dynamics of the western spruce budworm, *For. Sci.* **29**:779–790 (birds and ants may play an important role in maintaining sparse populations)

Crawford, H. S., Jr., and Jennings, D. T., 1982, Relationship of birds and spruce budworms—Literature review and annotated bibliography. U. S. Forest Service, Bibliography and Literature of Agriculture No. 23, 38 pp.

Embree, D. G., 1966, The role of introduced parasites in the control of the winter moth in Nova Scotia, *Can. Entomol.* **98**:1159–1167.

Harper, J. D., 1974, Forest insect control with *Bacillus thuringiensis*—Survey of current knowledge, University Printing Service, Auburn University, Alabama, 64 pp.

Holling, C. S., 1959, The components of predation revealed by a study of small-mammal predation of the European pine sawfly, *Can. Entomol.* **91**:293–320 (a key paper in developing ideas on functional and numerical responses)

Kroll, J. C., Conner, R. N., and Fleet, R. R., 1980, Woodpeckers and the southern pine beetle, U. S. Forest Service, Agriculture Handbook No. 564, 23 pp.

Miller, C. A., 1960, The interaction of the spruce budworm, *Choristoneura fumiferana* (Clem.), and the parasite *Glypta fumiferanae* (Vier.), *Can. Entomol.* **92**:839–850.

Mook, L. J., and Marshall, H. G. W., 1965, Digestion of spruce budworm larvae and pupae in the olive-backed thrush, *Hylochichla ustulata swainsoni* (Tschudi), *Can. Entomol.* **97**:1144–1149.

Moore, G. E., 1972, Southern pine beetle mortality in North Carolina caused by parasites and predators, *Environ. Entomol.* **1**:58–65.

Morris, R. F., 1959, Single-factor analysis in population dynamics, *Ecology* **40**:580–587.

Otvos, I. S., 1965, Studies on avian predators of *Dendroctonus brevicomis* LeConte (Coleoptera, Scolytidae) with special reference to Picidae, *Can. Entomol.* **97**:1184–1199.

Purrini, K., 1981, On sporozoan (Protozoa) disease agents of *Hylobius abietis* L. (Curculionidae, Coleoptera), *Proc. XVII IUFRO World Congr.* **2**:493–499.

Smith, H. R., and Campbell, R. W., 1977, Woodland mammals and the gypsy moth, *Am. For.* **84**, 22–25.

Smith, H. R., and Lautenschlager, R. A., 1981, Gypsy moth predators, in: The Gypsy Moth: Research Towards Integrated Pest Management (C. C. Doane and M. L. McManus, eds.), pp. 96–125, USDA Forest Service Technical Bulletin 1584.

Takekawa, J. Y. and Garton, E. O., 1984. How much is an evening grosbeak worth? *J. For.* **82**:126–128 (calculated value of this bird on the basis of what it would cost to achieve similar budworm mortality by spraying).

Thompson, C. G., and Hughes, K. M., 1977, Population management: Microbial control, in: The Douglas-Fir Tussock Moth: A synthesis (M. H. Brookes, R. W. Stark, and R. W. Campbell, eds.), pp. 133–140, U. S. Forest Service Technical Bulletin 1585.

Youngs, L. C., 1983, Predaceous ants in biological control of insect pests of North American forests, *Bull. Entomol. Soc. Am.* **29**(4), 47–50.

PART II: ECOLOGICAL EXERCISES

II.1. Pine bark beetles, *Dendroctonus ponderosae, D. frontalis,* or *D. brevicomis,* can only attack weakened pines when their populations are sparse, but large populations can overwhelm the resistance of relatively vigorous trees. Evaluate the equilibrium states for a pine bark beetle population in a stand of moderate vigor under the assumptions that (1) individual birth rates do not change with population density; (2) individual death rates reflect the availability of host material per beetle; and (3) the quantity of severely weakened trees remains constant with time.

    a. Draw an interaction diagram similar to that in Figs. 3.3, 3.4, or 3.7 showing the feedback structure of the bark beetle–pine interaction and identify (+) and (−) feedback loops.

    b. Draw a population diagram similar to Fig. 4.1 showing how the bark beetle interacts with its food supply and how environmental disturbances influence this interaction.

    c. Draw the individual density-dependent birth and death curves for the bark beetle, i.e., how birth and death rates change with population density as in Figs. 4.2 and 4.4. Remember that the death rate should increase with population density when food is constant because the food available per individual will decline. Also consider that the death rate should decline with density when populations become dense enough to overcome vigorous trees because the food available per individual will increase. Finally the death rate, or more probably the emigration rate, should rise again at very high densities as the pine stand is destroyed and food availability per beetle becomes very low.

    d. Identify the potential equilibrium points and discuss the stability characteristics.

    e. Repeat the analysis with a stand of higher and lower overall vigor but otherwise under the same set of assumptions. Remember that larger beetle populations will be required to overwhelm more vigorous trees. How does stand vigor affect the outbreak threshold?

    f. From the previous analysis draw the threshold function for lodgepole pine (e.g., see Fig. 4.4d).

    g. What kind of outbreaks are exhibited by these beetles?

II.2. In his studies on the population dynamics of the gypsy moth, *Lymantria dispar,* a defoliator of deciduous trees (oaks are the preferred hosts), Campbell (1974) drew the following conclusions concerning the effects of population density on the individual birth and death rates:

    • The individual birth rate remains more or less constant until the insect reaches very high densities where food shortages cause decreased fecundity in malnourished females.

    • The individual death rate curve was much more complex. In sparse gypsy moth populations, the death rate increases with population density because birds, mice and parasitic insects respond by concentrating on gypsy moth larvae and pupae as their abundance increases. As densities get moderately high, however, the birds and mice become satiated and the parasites become less effective. This causes the death rate to decline with population density. Then, as the moth population becomes very dense, the death rate again rises rapidly with population density because caterpillars starve from lack of food and are killed by viral pathogens (see Campbell, 1974, Fig. 20).

    a. Draw a flow diagram similar to that in Fig. 4.1 showing the interactions between the gypsy moth and its food supply (the foliage of deciduous trees) as well as its natural enemies (birds, mice, insect parasites, and virus).

    b. Describe how the favorability of the environment for an individual gypsy moth changes with the density of its population.

    c. Draw the individual birth and death rate curves as they change in relation to population density, and identify the potential equilibrium points. Explain whether the equilibrium points are stable or unstable and classify the gypsy moth outbreak type. Reference to Campbell and Sloan (1978) will aid in classifying the insect's population dynamics.

    d. What effects would you expect if the maximum birth rate were increased or decreased because of changes in the basic physical environment?

e. What effects might occur if the environment was made less favorable for birds and mice or if hiding places were abundant for moth larvae and pupae? For instance, Campbell et al. (1976) found pupae to be protected from predation by bark flaps, signs nailed to trees, abandoned car bodies, and other human debris. How might this affect the dynamics of gypsy moth populations along the edges of forests where junk is abundant?

II.3 Mattson (1980) estimated the total number of red pine cones produced per acre in a seed-producing area and also estimated the total number of cone beetles present. Read this paper and then perform the following analysis:

a. Plot on a logarithmic scale the mean number of cones and cone beetles per year against time. What is the equilibrium density for cones and beetles? How do beetle populations respond to the abundance of food?

b. Draw a flow diagram, similar to that in Fig. 4.1, of the interactions between the beetle and its food supply. What kinds of feedback loops are likely to be operating?

c. Calculate the individual growth rate of the population for each year using the equation

$$r_t = \log_e(N_{t+1}/N_t) = \log_e(b - d + i - e + 1)$$

where N_t is the number of beetles per acre in year t and N_{t+1} is the number in the next year. Graph these values against the number of cones available per beetle in year t, that is,

$$F_t = C_t/N_t$$

where F_t is the favorability of the environment for the beetle in terms of cone availability per beetle, and C_t is the abundance of cones (basic environment). What does this graph tell you about the individual birth, death, and migration rates of the insect? If the cone supply were artificially maintained at a constant level of 30,000 per hectare, what would be the equilibrium density of the beetle population?

d. Calculate and draw the equilibrium beetle density as a function of cone crop size similar to that in Fig. 4.2d. What does this tell you about the outbreak behavior of the cone beetle population? What is the type of outbreak?

II.4. An experiment was set up to measure the rate of prey capture by two predators over a 1-day period. The capture rate was found to be related to prey population density in the following manner:

Prey density	100	200	300	400	500	600	800	1000		

(Prey caught/predator per day)

Predator 1	24	33	38	39	40	40	40	40
Predator 2	2	9	27	37	40	40	40	40

a. Plot the functional responses of these two predators on graph paper. Describe their differences and the probable causes of these differences.

b. Suppose that we measured the daily individual birth and death rates of the prey species in the absence of the predators and found them to be $b = 2.48$, $d_o = 2.34$, and to be independent of population density. Calculate the individual death rate of the prey attributable to the feeding of a single individual of predator #1 using the relationship $d_p = P \cdot K/N$, where P is the density of predator #1, and K is the number of prey killed at density N. Plot the prey birth and death rates over the range of population density observed on the same graph. How does this predator affect the prey population?

   c. Repeat the above exercise with a single individual of predator #2. Can this single predator regulate the density of its prey?

   d. Repeat the above exercise with two individuals of predator #2. Can these predators regulate the prey population? At what density is the prey regulated? At what density can the prey escape from regulation?

II.5 Analyze the interactive structure and potential behavior of the insect chosen for your special project.

   a. Describe, if possible, the means by which host plants defend themselves against this insect and how the insect population can influence the structure of forest communities.

   b. Draw an interaction diagram similar to that in Figs. 3.3, 3.4, and 3.7 showing interactions among the insect, its host plant(s), and its important natural enemies. Identify potential feedback loops and their relative significance in stabilizing or destabilizing this system.

   c. Draw a population diagram using Fig. 4.1 as a model. What does the drawing tell you about the stability of this system?

   d. Deduce the individual density-dependent birth and death curves; identify potential equilibria and their probable stability properties.

   e. Identify the class of outbreak behavior exhibited by this insect.

   f. What basic environmental factors may stabilize or destabilize this system?

MANAGEMENT

RISK CLASSES FOR PONDEROSA PINE

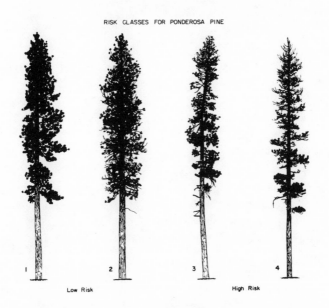

Low Risk   High Risk

"While the sluggard turns to the ant for his model of industry, the forester turns to the bark beetle to learn how to manage a forest."

F. P. Keen, *Pan-Pacific Entomologist,*
Vol. 22, 1946.

Drawing courtesy of USDA Forest Service.

CHAPTER 6

MONITORING AND FORECASTING INSECT OUTBREAKS

Most forestry organizations periodically monitor their lands in order to detect incipient pest outbreaks and to determine damage levels or changes in the boundaries of ongoing infestations. These surveys may be designed to monitor large areas of forest (extensive surveys) or to examine particular infestations in considerable detail (intensive surveys). In addition to providing information on current damage levels, survey data may be used to forecast damage trends.

6.1. EXTENSIVE SURVEYS

Extensive surveys are designed to monitor large areas of forest with diverse ownerships and are usually planned, coordinated, and executed by government agencies. For example, the Forest Pest Management Division of the United States Forest Service is responsible for surveillance activities on all federal lands; it also cooperates with other federal and state agencies in surveying other ownerships. Each year the state and federal governments publish cooperative reports on the major insect and disease problems of each region; the Forest Service also publishes a summary report for the entire country. Likewise, the Forest Protection Branch of the Canadian Forestry Service organizes a highly efficient surveillance program and publishes an annual report for the whole of Canada. Large private companies may also survey their lands more frequently if they feel the need for closer monitoring.

Aircraft are used most frequently for extensive damage surveys because they can cover large areas quickly and relatively cheaply. These planned systematic surveys are carried out at the time of year when insect damage is most easily seen from the air. Aircraft are usually flown at a height of about 1000 ft along parallel lines or, in steep terrain, along contour lines. One or two observers in the plane then sketch the boundaries of visible insect damage on a map or aerial photograph of the area (Fig. 6.1.). Highly skilled observers are often able to recognize tree species and damage symptoms and to make a good guess at the species of insect involved. Tree and insect species are then marked on the map in color code or with alphanumeric identification labels (Fig. 6.1). When there is doubt about the aerial identification, ground checks should be carried out by experienced entomologists.

Because aerial surveys have to be flow when the symptoms of insect damage are most visible, careful planning and timing is essential. Flights are often scheduled to coincide with the visible symptoms of several important insect species. Frequently, however, more than one flight per year is needed because the symptoms of different species do not coincide. In warmer regions, in which insect populations may undergo more than one generation per year, sequential surveys may also be necessary throughout the year.

In addition to organized and systematic aerial surveys, many forest managers train their ground crews in insect and disease detection and recognition. Trained personnel, who spend most of their time in the field, are an invaluable aid in pest surveillance, often detecting incipient pest problems before they become visible from the air. In fact, many insect problems cannot be seen from the air at all, such as tip and shoot insects and cone insects. One of the most efficient ground surveillance systems is that practiced by the Canadian Forestry Service, which employs forest biology rangers to carry out detailed examinations of insect infestations and to collect specimens for shipment to specialists for rearing and identification. Ground examinations not only provide information on the pest species involved but may give clues to population trends and the effectiveness of natural enemies and diseases in checking outbreaks. In some forest regions with good road access and large ground crews, ground surveys may be the only surveillance method used. In central Europe, for example, foresters get to see every tree in the forest once every few years.

Other methods are sometimes employed for detecting insect activity. For example, vantage points such as fire towers are occasionally used to obtain an overview of a specific area, while road surveys may be undertaken, particularly when road systems run along ridges or otherwise provide an overview of the country.

Whatever supplemental techniques are used for monitoring insect infestations over large areas, aerial surveys remain the backbone of the surveillance system in those countries with extensive forest resources. These surveys provide forest managers with an extensive overview of insect damage on their lands. Aerial surveys, however, are notoriously inaccurate and are often behind the times. For example, infestations recorded by aerial observers rarely conform exactly to ground coordinates, and the intensity of the infestation or the number of trees actually infested may be off by an order of magnitude. In addition, aerial surveys are often in error because damage to trees, rather than actual insect

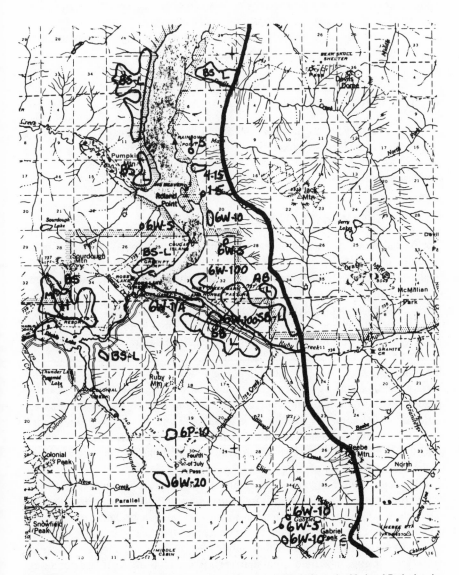

FIGURE 6.1. A portion of the 1978 Forest Insect Survey of North Cascades National Park showing infestation boundaries of the Douglas-fir beetle (1), fir engraver beetle (4), mountain pine beetle in ponderosa (6P) and white (6W) pines, balsam woolly adelgid (AB), and western spruce budworm (BS). The species code is followed by an estimate of damage intensity (L, light; M, moderate; H, heavy) or the number of trees killed. (USDA Forest Service, Forest Insect Survey Map.)

populations, are recorded; i.e., the damage may have been caused by past populations that are no longer present in the area. To obtain more accurate information, aerial surveys are often backed up by intensive ground surveys designed to determine actual damage, the size of the insect population, and the potential for future damage.

6.2. INTENSIVE SURVEYS

Intensive surveys are designed to obtain accurate estimates of damage or insect population levels by counting infested trees, identifying levels of tree damage, and/or counting the insects and their natural enemies. These surveys may be carried out from the air or by ground crews.

6.2.1. Intensive Aerial Surveys

Aerial photography is the cheapest and quickest way to obtain accurate estimates of insect damage to the forest. This is particularly true when bark beetles are involved because the dead trees have characteristic bright red foliage, which is easily visible on color or infrared photographs.

A variety of photographic equipment, film types, and aircraft have been tested for aerial damage appraisal, ranging from 35-mm stereoscopic color photography from small low-flying aircraft to sophisticated infrared panoramic photography from high-flying U2 aircraft and space satellites. Different techniques and film types have their own advantages and disadvantages, but all seem to be effective and reasonably accurate. The choice of technique is therefore, usually decided by the equipment and expertise available in a given region. In general, the accuracy of estimating damage from aerial photographs improves with larger scales (lower altitudes), stereoscopic viewing, and the experience of the photointerpreter. When high-flying aircraft are used, infrared photography is necessary to eliminate the effects of haze.

The most commonly used procedure for aerial damage appraisal in the United States is the so-called *multistage sampling technique*. The first stage is an extensive survey in which the general infestation boundaries and a rough estimate of damage intensity are sketched on a map of the area (e.g., see Fig. 6.1). Using this sketchmap, the area is then subdivided into discrete damage classes or strata (e.g., heavy, moderate, light damage). Photo plots are then laid out along parallel flight lines with the objective of obtaining an adequate sample size from each infestation stratum (Fig. 6.2). After the photographs have been taken, a random subsample of these plots is usually selected for "ground truth" data acquisition. Small plots are randomly selected from aerial photographs, located on topographic maps, and then visited by ground crews who count the infested trees. These ground truth data are then used to check the precision of the aerial photographic interpretation and to adjust the counts accordingly. Standard statis-

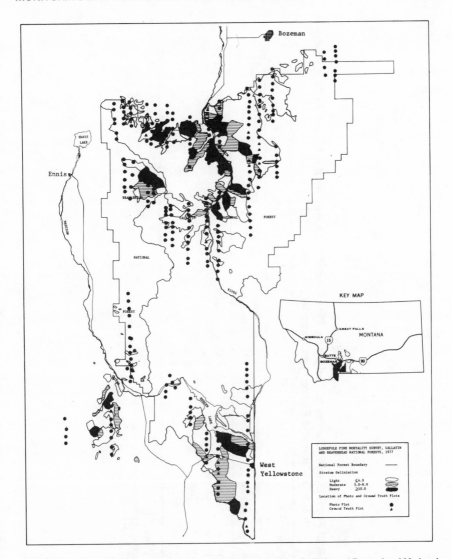

FIGURE 6.2. Photoplot layout for an intensive aerial survey of the Gallatin and Beaverhead National Forests to measure lodgepole pine mortality caused by the mountain pine beetle. (Reproduced from Bennett and Bousfield, 1979).

tical routines are then used to estimate the total number of trees infested and the total volume lost in each damage stratum (Table 6.1).

6.2.2. Intensive Ground Surveys

Ground surveys are used when detailed information is required on the damage caused by a specific insect, or a group of different species, when these

TABLE 6.1
Results of a Multistage Aerial Photographic Survey of Lodgepole Pine Mortality Caused by the Mountain Pine Beetle on the Gallatin and Beaverhead National Forests in 1977[a]

	From sketchmaps					From aerial photo interpretation			
Stratum	Infested trees/acre	Acres infested	Photo plots (40 acre)	Ground plots (2.5 acre)	R^{2b}	Estimated trees killed	Standard error (%)	Estimated volume lost[c]	Standard error (%)
1	≤4.9	179,325	94	24	0.82	467,938	7.2	10,370	15.5
2	5.0–9.9	52,231	24	16	0.94	462,406	7.8	8,549	17.6
3	≥10	38,649	26	16	0.94	340,273	7.7	5,317	18.2
Totals		270,205	144	56	0.90	1,270,617	4.4	24,236	9.9

[a]See Fig. 6.2 and references cited therein.
[b]R^2 coefficient of determination between photo interpretation and ground plots.
[c]Millions of cubic feet.

FIGURE 6.3. Sampling procedures for forest insect and disease surveys. First a small (1/300 acre) fixed plot is set up around the sampling point and small trees (e.g., less than 5 in. DBH), and sometimes other vegetation (annuals and shrubs) are measured. (a) Then, in the fixed plot design a larger (e.g., 1/10 acre) fixed plot is laid out and all larger trees measured. (b) In the variable plot design a forester's prism is used to determine which trees to measure. This method samples larger trees, which occur with less frequency, with greater sampling probability than small ones. (c) The forester's prism is held at arms length and the tree observed through it. This tree would be sampled because the prism image is not deflected beyond the true diameter of the tree. (a and b redrawn from Bousfield, 1980). (●) sampled tree; (○) tree not sampled.

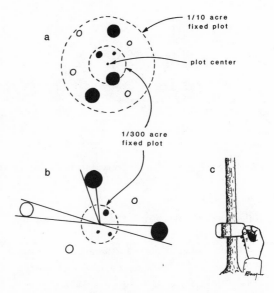

data cannot be obtained by aerial surveys. Once again, stratified random sampling procedures are usually employed, with infestation strata delineated from extensive aerial sketchmaps or photographs. Within each infestation strata (e.g., heavy, medium, and light infestations), a group of stands is selected at random, and ground crews lay out a series of sampling points through each stand. For example, sampling points may be placed at five-chain* intervals along transects 10 chains apart to give 20 sampling points in a 100-acre stand. As the number of sampling points is usually fixed to provide certain levels of precision, sampling intensity will necessarily reflect the size of the stand being sampled; i.e., small stands will have more sampling points per unit area than large ones.

At each sampling point, trees are selected for measurement using either a *fixed* or *variable* plot design (Fig. 6.3). Both methods are designed to assign a higher sampling probability to large trees, which occur at lower frequencies than small ones. Sample trees are then measured for various attributes, such as diameter at breast height (DBH), height, crown length and width, growth rates and age from increment cores, and the presence of insects or insect damage. Insect damage may be classified into defoliation categories (Fig. 6.4), into dead, unsuccessfully attacked and unattacked trees (bark beetles), top kill (defoliators and bark beetles), stem deformities (tip insects), and so on. In addition, data may be

---

*A chain is a unit of measure used in forest-land surveying; 1 chain = 66 ft or 20 m.

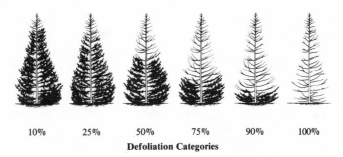

10%   25%   50%   75%   90%   100%

Defoliation Categories

FIGURE 6.4. Classification of trees according to percentage of the crown defoliated by the Douglas-fir tussock moth. (From Wickman, 1979.)

collected on the occurrence of seedlings, saplings, shrubs, herbs, and forbs using a small fixed plot or thrown quadrants.

Data gathered during ground surveys are usually coded in computer-readable form and then analyzed and summarized by special computer programs. Data summaries usually include stand inventory information, numbers of trees damaged, and volume or board foot losses per unit area of forest (e.g., per acre of hectare).

6.2.3. Sampling Insect Populations

In some instances, particularly when predictions of future damage are required, insect populations and their natural enemies may be measured directly. This activity entails even more intensive sampling efforts.

Because insect populations are rarely distributed uniformly or randomly throughout infested trees, stratified sampling is usually employed to represent the true variation in the population. Stratification should be based on information on the density distribution of the insect population within infested trees. For example, research on the distribution of bark beetles has shown that the major sources of variation are with vertical height in the tree. Hence, the infested height of the tree is usually divided into sampling strata (Fig. 6.5a). In other cases, circumferential (around the tree) or radial (from inside to outside) variations may be significant, necessitating stratification in these dimensions (Fig. 6.5b).

Optimal sampling designs are those that provide a good estimate of the population mean density with the least number of samples. Of course, decisions on sampling design always involve tradeoffs between effort and precision. Thus, some designs will sample more strata and provide greater precision at greater cost than others designed to minimize costs. One commonly used procedure for determining optimal sampling design is to compare the variation within and between trees with their respective sampling costs. The optimal number of sampling strata per tree, n_s, is then given by

$$n_s = \sqrt{\frac{s_s^2 \cdot C_p}{s_p^2 \cdot C_s}}$$ (6.1)

where s_s^2 is the variance within the tree, C_s is the cost of obtaining a sample, s_p^2 is the variation between trees, and C_p is the cost of moving from one tree to another.

The mechanics of sampling varies greatly with the kind of insect being sampled. Bark beetle populations are usually sampled by cutting out an area of bark with a hand or rotary saw, defoliators by clipping or beating a known quantity of foliage (Fig. 6.6), cone insects by climbing the tree or shooting down cone-bearing branches, tip insects by clipping terminals, and so on.

Once the samples have been taken, the insect pests and their developmental stages are identified, or preserved for later identification, and counted. In some cases, parasitic and predaceous organisms, as well as insects infected with pathogenic microorganisms, are also noted. Because many forest insects live inside bark, wood, cones, seeds, or terminals, counting the insects may become very laborious. This task can be made easier by the use of X-ray films (e.g., see Fig. 5.8).

Data from insect samples may be summarized as the average density of insects per unit area of habitat (e.g., per square meter of foliage, bark) or in absolute terms as the number per unit area of forest (per hectare). These data may then be converted into damage (e.g., percentage defoliation, probability of tree

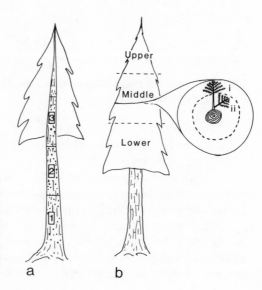

FIGURE 6.5. Optimal sampling stratifications. (a) Removing half-square-foot bark sections for sampling the fir engraver beetle. (From Berryman, 1968). (b) Clipping an outside (i) and inside (ii) branch from three crown strata for sampling Douglas-fir tussock moth larvae. (From Mason, 1979.)

a   b

FIGURE 6.6. Sampling Douglas-fir tussock moth populations by clipping (a) and beating (b) the foliage. (From Mason, 1979.)

death, deformity) if the functional relationships between insect population density and damage are known.

In cases in which several samples have been taken throughout the life cycle of the insect, the data are often presented as *life tables* (Table 6.2). Life tables illustrate the changes in the insect population through its life cycle and the effect of mortality agents acting on the various life stages. They enable one to see at a glance which developmental stages are most vulnerable to mortality agents, which mortality agents have the greatest impact on the population, and the rate of increase, or explosive potential of the population over its life cycle.

FIGURE 6.6. (*continued*)

6.3. DAMAGE FORECASTS

Data obtained during forest insect surveys or insect population sampling are often employed to forecast future insect populations or damage levels. Accurate forecasts are extremely useful to the forest manager for they provide time to plan management actions.

6.3.1. Trend Indexes

The simplest and most widely used forecasting method is the so-called *trend index*. This is a measurement of the trend of an insect population, or the damage resulting from an insect infestation, from one year, or generation, to the next. The trend index is specified by

$$I = N_t/N_{t-1} \qquad (6.2)$$

where N_t is the density of the population at a particular stage of development in the current year or generation, and N_{t-1} is the density of the same stage in the previous one. In effect, the trend index is a measure of the rate of population

TABLE 6.2
Life Table for the Lodgepole Needle Miner, 1954–1956 Generation[a]

x—Age interval	1_x—No. alive at beginning of x	d_xF—Factor responsible for d_x	d_x—No. dying during x	$100q_x$—d_x as percentage of 1_x
X_1—Eggs	4700	Needle drop Larval dispersion Predators	3586	76.30
X_2—Instars I and II	1114	Climate—late fall and early winter temperatures	409	36.68
Extra—Dec. 14, 1954	705	Climate—late winter temperatures	219	31.06
		Spring kill	70	9.93
			289	40.99
X_3 } Instars III and IV X_4 } July 1, 1955	416	Winter mortality	134	32.18
		Parasitism	142	34.09
		Spring kill	9	2.21
		Bird predation	1	0.30
		Unknown	7	1.68
			293	70.46
X_5—Instars IV and V	123	Parasites	18	14.55
X_6—Pupae	105	Unknown	26	24.76
		Parasites	<1	0.45
			26	25.21
Emerged	79			
Moths SR 48:52	M F 38 41			
Generation			4621	98.32

[a]From Stark (1958).

change over a generation or, as we saw in Chapter 4, of the net rate of individual birth, death, immigration, and emigration; i.e.,

$$I = N_t/N_{t-1} = b - d + i - e + 1 = \exp(r) \qquad (6.3)$$

If we assume that the trend index remains the same over the next time period, the forecasted population density, N_{t+1}, is given by

$$N_{t+1} = I \cdot N_t \qquad (6.4)$$

The trend index also provides a qualitative estimate of population change, because populations are expected to increase when $I > 1$, to decrease when $I < 1$, and to remain unchanged when $I = 1$.

Trend indexes can also be calculated from data on damage intensity (dead trees per hectare or percentage defoliation) or from the area of land infested (hectares or acres) from one time to the next. Thus, we can estimate three different trends:

1. *Population trend:*
   $I_n = N_t/N_{t-1}$
   where N_t = population density at time t
2. *Damage trend:*
   $I_d = D_t/D_{t-1}$
   where D_t = damage intensity at time t
3. *Spread trend:*
   $I_a = A_t/A_{t-1}$
   where A_t = area infested at time t.

The problem with forecasts based on the trend index is that the index is assumed constant over the forecasting period. In fact, it is a highly variable statistic influenced by many factors that affect insect fecundity, survival, and movement. If information is available on the expected fecundity, mortality, and dispersal of the population during the forecasting period, more accurate forecasts are possible. It is here that life tables describing the schedule of births, mortality, and dispersal through the life cycle of the insect (Table 6.2) can be extremely helpful.

Special analytical techniques are available for determining which of the components of the life table act as *key factors* in determining population variations. One method is to plot the trend index, or preferably the exponential growth rate of the population, that is, $r = \log_e(I)$ [Eq. (6.3)], against the variable in question (Fig. 6.7). Key factors will be those that explain most of the variation in population growth rates. In the case of the black-headed budworm (Fig. 6.7), the individual population growth rate is strongly correlated with the probability of an individual being parasitized (the correlation coefficient is 0.93), suggesting that this variable may have good predictive power. We see, for example, that budworm populations always increase ($I > 1$) when less than 32% of the larvae are parasitized in the current generation. The regression equation (Fig. 6.7) can be rearranged to provide a forecasting model; i.e.,

$$I = N_{t+1}/N_t = \exp(a + b \cdot P_t)$$
$$N_{t+1} = N_t \cdot \exp(a + b \cdot P_t) \qquad (6.5)$$

where P_t is the probability of parasitism in year t, and a and b are the regression coefficients, e.g., $a = 1.48$, $b = -4.1$, for the black-headed budworm.

Although forecasting approaches that incorporate the effect of major mortality factors [Eq. (6.5)] usually produce more reliable predictions than does the simple trend index [Eq. (6.4)], they may still suffer from the omission of factors that have significant effects on population change. More reliable forecasts may

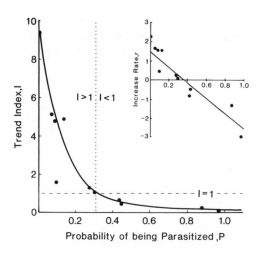

FIGURE 6.7. Relationship between the probability of a black-headed budworm, *Acleris variana,* larva being parasitized and the population trend index ($I = N_t/N_{t-1}$), and the individual rate of increase ($r = \log_e I$). Note that the population declines ($I < 1$ or $r < 0$) when more than 32% of the budworms are parasitized: curve fit, $r = 1.48 - 4.1P$; $r^2 = 0.86$). (Data from Morris, 1959).

be possible if the models incorporate multiple predictive variables. For instance, the trend index can be specified in general terms by

$$I = F(X_1, X_2, X_3, \ldots)\cdot S(Y_1, Y_2, Y_3, \ldots) \pm M \qquad (6.6)$$

where F is the average fecundity (birth rate) per individual expressed as a function of a set of environmental variables, X_i; S is the probability of individuals surviving from egg to adult expressed as a function of another set of environmental variables, Y_j; and M is the net migration into the area. Obviously, the construction of such complex forecasting models requires a large number of data and a considerable investment in research.

6.3.2. Sequential Sampling

Sequential sampling was originally developed for quality control of industrial products, that is, for determining if a batch of products meets certain quality requirements. The method has been used for sampling forest insect populations when interest is centered on qualitative predictions; that is, if we are interested in determining whether the population or damage is increasing or decreasing, rather than in the precise levels of insect population or damage. Sequential sampling is extremely cost effective because it reduces the number of samples needed to arrive at a decision to a bare minimum.

The design of a sequential sampling scheme requires a considerable amount of biological information and involves some rather formidable mathematics. The result, however, is a simple decision graph (Fig. 6.8), which is used in the following manner: A minimum number of samples is usually taken (20 in our example) and the total number of insects summed for all samples. If the cumula-

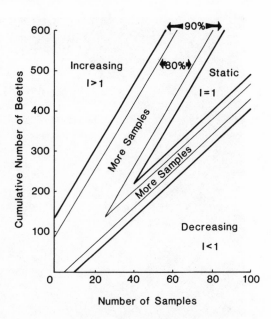

FIGURE 6.8. Sequential sampling plan for the mountain pine beetle infesting ponderosa pine stands in the Black Hills. (After Knight, 1960.) Two 6×6-inch bark samples are removed from the North and South sides (1 from each side) of each infested tree (at breast height). Ten trees are sampled first, and sampling is then continued until the cumulative beetle count falls within a decision zone or sampling is discontinued (at 40 trees). Decisions to stop sampling can be made at the 80 or 90% levels of confidence.

tive count falls within one of the decision zones, then sampling is terminated and a prediction is made on the status of the population. If the datum falls within the zone of indecision, sampling continues until the cumulative count falls in a decision zone, or sampling is terminated for other reasons (an upper sampling limit is usually set for practical purposes).

Sequential sampling is a useful technique when quick and relatively cheap estimates are required of the general status of pest populations. For example, whether populations or damage are increasing, decreasing or static, or if populations are at outbreak, suboutbreak, or nonoutbreak levels. For this reason, sequential sampling is frequently used to determine the spatial boundaries of insect outbreaks.

6.3.3. Pheromone Trap Forecasts

A recent innovation is the use of insect pheromones for forecasting population trends. The gypsy moth sex pheromone, for example, is employed throughout the United States and Canada to trace the spread of this exotic pest into new areas. In the Pacific Northwest, the sex pheromone of the Douglas-fir tussock moth is used to monitor moth populations continuously during their low-density periods (Fig. 6.9). Pheromones are so sensitive that they will attract male moths when they are very scarce and cannot be observed by any other means. For this reason they are most useful for detecting the presence of exotic species (gypsy moth) or as an early warning of incipient outbreaks (Douglas-fir tussock moth;

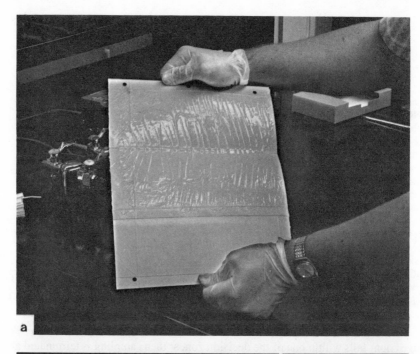

FIGURE 6.9. Assembly and placement of the "milk carton" pheromone trap used for detection of Douglas-fir tussock moth population increases. (a) Unfolded trap showing sticky inner surface. (b) Polyvinyl plastic pheromone dispensers being placed into the folded trap.

c

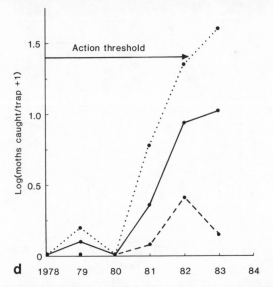

d

(c) Deployed trap containing male tussock moths. (d) Captures of male tussock moths at two locations in Latah County northern Idaho. (. . .) Mineral Mountain; (- - -) East Dennis; (—) the average capture for six plots. When moth captures exceed the action threshold larval sampling is initiated to obtain more accurate population estimates. (Data from R. L. Livingston, Idaho Department of Lands; photographs from Daterman *et al.*, 1979.)

see Fig. 6.9d). Trap captures from one year to the next can also be used to calculate a trend index [Eq. (6.2)].

6.3.4. Forecasting Models

The degree of damage expected during an insect outbreak can sometimes be estimated by regression models that relate damage, in terms of tree mortality, percentage defoliation, tree deformity, and so forth, to certain site and/or stand variables. They are discussed in more detail in Chapter 7.

Complex simulation models may also be available for projecting the effects of insect outbreaks into the future. In the Western United States, for example, outbreak models have been constructed for the mountain pine beetle, Douglas-fir tussock moth, and western budworm. These models operate on stand and site inventory data, together with sample estimates of pest, parasites, predators, and disease abundance and then calculate the expected population densities and damage levels over the course of the outbreak (Fig. 6.10). Insect outbreak models can also be linked to stand growth models to project the effects of insect

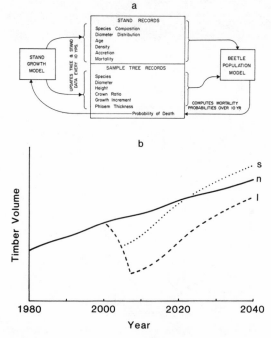

FIGURE 6.10. (a) Schematic of a model for projecting timber losses due to the mountain pine beetle from inventory into the future. (After Berryman, 1982.) (b) Projected timber volumes from inventory in the presence (l) and absence (n) of a mountain pine beetle outbreak and when half the killed trees are salvaged (s). (Adapted from Crookston and Stark, 1984.)

infestations over the entire rotation period. Interactive stand growth–insect out-break models are often used as aids to management decision-making and are discussed in more detail in Chapter 10.

REFERENCES AND SELECTED READINGS

Aldrich, R. C., and Drooz, A. T., 1965, Estimating fraser fir mortality and balsam woolly aphid infestation trend using aerial color photography, *For. Sci.* **13:**300–313. (intensive aerial survey)

Avery, T. E., 1966, Forester's guide to aerial photo interpretation, U.S. Forest Service, Agricultural Handbook No. 308, 40 pp.

Bennett, D. D., and Bousfield, W. E., 1979, A pilot survey to measure annual mortality caused by the mountain pine beetle in lodgepole pine on the Beaverhead and Gallatin National Forests, 1978, U.S. Forest Service, Northern Region State and Private Forestry Report No. 79-20, 13 pp. (intensive aerial survey)

Berryman, A. A., 1968, Development of sampling techniques and life tables for the fir engraver, *Scolytus ventralis* (Coleoptera: Scolytidae), *Can. Entomol.* **100:**1138–1147. (optimal sampling procedures)

Berryman, A. A., 1982, Mountain pine beetle outbreaks in Rocky Mountain lodgepole pine forests. *J. For.* **80:**410–413.

Berryman, A. A., and Stark, R. W., 1962, Radiography in forest entomology, *Ann. Entomol. Soc. Am.* **55:**454–466. (use of X-rays for sampling cryptic insects)

Bousfield, W. E., 1980, R-1 forest insect and disease damage survey system, U.S. Forest Service Service, Northern Region State and Private Forestry, Report No. 79-2, Missoula, 23 pp. (intensive ground survey)

Ciesla, W. M., Allison, R. A., and Weber, F. P., 1982. Panoramic aerial photography in forest pest management, *Photogram. Eng. Remote Sensing* **48:**719–723.

Crookston, N. L., and Stark, R. W., 1985, Forest–bark beetle interactions: Stand dynamics and Prognoses, pp. 81–103, in: *Integrated Pest Management in Pine–Bark Beetle Ecosystems* (W. E. Waters, R. W. Stark, and D. L. Wood, eds.), Wiley (Interscience), New York.

Daterman, G. E., Livingston, R. L., Wenz, J. M., and Sower, L. L., 1979, How to use pheromone traps to determine outbreak potentials, U.S. Forest Service, Agricultural Handbook No. 546, 11 pp. (for Douglas-fir tussock moth)

Dottavio, L. C., and Williams, D. L., 1983, Satellite technology: An improved means of monitoring forest insect defoliation, *J. For.* **81:**30–34.

Environment Canada, Forest insect and disease survey, Annual Reports, Canadian Forestry Service, Ottawa.

Environment Canada, Forest insect and disease conditions in Canada 1980 (and annually), Canadian Forestry Service, Ottawa.

Gimbarzevsky, P., 1984, Remote sensing in forest damage detection and appraisal: Selected anno-tated bibliography, Canadian Forestry Service Information Report BC-X-253.

Heller, R. C., Coyne, J. F., and Bean, J. L. 1955, Airplanes increase effectiveness of southern pine beetle surveys, *J. For.* **53:**483–485.

Heller, R. C., and Schmiege, D. C., 1962, Aerial survey techniques for the spruce budworm in the Lake States, *J. For.* **60:**525–532.

Klein, W. H., Bennett, D. D., and Young, R. W., 1980, Evaluation of panoramic reconnaissance aerial photography for measuring annual mortality of lodgepole pine caused by the mountain pine beetle, U.S. Forest Service, Northern Region, Report No. 80-2, 21 pp. (intensive aerial survey)

Knight, F. B., 1960, Sequential sampling of Black Hills beetle populations, U.S. Forest Service, Rocky Mountain Forest and Range Experiment Station, Research Note 48.

Mason, R. R., 1979, How to sample Douglas-fir tussock moth larvae. U.S. Forest Service, Agricultural Handbook No. 547, 15 pp.

Monserud, R. A., and Crookston, N. L., 1982. A user's guide to the combined stand prognosis and Douglas-fir tussock moth outbreak model, U.S. Forest Service General Technical Report INT-27, 49 pp.

Morris, R. F., 1955, The development of sampling techniques for forest insect defoliators, with particular reference to the spruce budworm, *Can. J. Zool.* **33:**225–294. (a classic treatise on sampling forest insects)

Morris, R. F., 1959, Single-factor analysis in population dynamics, *Ecology* **40:**580–588. (method for identification of major mortality, or key factors; also covered by Southwood, below).

Sippell, W. L., 1983, Continuous evaluation and reporting of a changing spruce budworm damage picture in Canada. Canadian Forestry Service, CANUSA, Quebec.

Southwood, T. R. E., 1966, *Ecological Methods with Particular Reference to the Study of Insect Populations,* 2nd ed., 1978. Chapman and Hall, London, 524 pp. (excellent review of sampling methods, life tables, key factors analysis, etc.)

Stark, R. W., 1958, Life tables for the lodgepole needle miner *Recurvaria starki* Free (Lepidoptera: Gelechiidae), *Proc. 10th Int. Congr. Entomol.* **4:**151–162.

USDA, 1981, Forest insect and disease conditions in the United States, 1979, U.S. Forest Service, General Technical Report WO-20, 91 pp. (published annually).

Waters, W. E., 1955, Sequential sampling in forest insect surveys, *For. Sci.* **1:**68–79 (sequential sampling methods)

Wear, J. F., Pope, R. B., and Lauterbach, P. G., 1964, Estimating beetle-killed Douglas-fir by aerial photo and field plots, *J. For.* **62:**309–315. (intensive aerial survey)

Wickman, B. E., 1979, How to estimate defoliation and predict tree damage, U.S. Forest Service Agricultural Handbook No. 550. (measuring Douglas-fir tussock moth defoliation)

CHAPTER 7

ASSESSING THE RISK OF INSECT OUTBREAKS

In the preceding chapter we discussed methods for monitoring ongoing forest insect infestations and for making short-term forecasts. These procedures address the manager's concern with current conditions—where damage is occurring at present, how much damage is being done, and whether this damage will increase or decrease in the near future. By contrast, risk assessment methods are intended to predict the likelihood of insect problems in the more distant future—where and under what conditions insect outbreaks are likely to occur in the future.

First let us clearly explain what we mean by the term risk. The literature on this subject is somewhat confusing, with such words as risk, hazard, susceptibility, and vulnerability sometimes being used for the same thing. To avoid this confusion, we shall restrict ourselves to the term risk and define it as follows: *Risk* is the *probability* of an insect outbreak occurring in a particular stand, watershed, or forest, or of a particular tree being severely damaged, under a given set of conditions. The concept of risk does not concern itself with the extent of damage but purely with the event itself—whether an outbreak will or will not occur or a tree will or will not die. Risk, therefore, has a common meaning in all walks of life—the risk of an automobile accident, contracting a disease, or suffering an insect outbreak or fire—whereas losses due to the event are usually measured after the fact (Chapter 6; see also Chapter 10).

Like any insurance company, forest managers are interested in finding association between risk and easily measured variables. If this can be done, risky situations can be avoided by controlling or avoiding the critical conditions. As we saw in Chapter 4, insect outbreaks are often associated with the basic properties of the environment. In environments that are favorable for the pest, or unfavorable for its natural enemies or host plant, we usually find consistently high pest densities or violent population cycles (see Fig. 4.12, right column).

Risk assessments, therefore, are often based on an analysis of basic environmental conditions, particularly of those variables associated with the physical *site*, the forest *stand*, and that act as *disturbances* to normal conditions (see Fig. 4.1).

Although risk assessments are usually designed to operate at the stand level, they can also be made on an individual tree basis. This latter approach is often used when trees are killed individually (bark beetles) and where the risk can be assessed as the probability of tree death.

7.1. RISK ASSESSMENT VARIABLES

7.1.1. Site Factors

The land on which a particular forest stand grows can be characterized by a set of physical parameters usually called site factors. These factors include soil structure, chemistry, and bedrock formation; topographic features such as elevation, slope, and aspect; and regional climatic and weather patterns. These characteristics play a large part in determining the kinds of vegetation that grow on a particular site.

The suitability of a particular site for plant growth depends on the levels of nutrients and water in the soil and on temperature. These features have an overriding influence on the plant community and its development toward the final *climax* state. In other words, they determine the *habitat type* of a particular piece of land. Because insects are dependent on the presence of their plant food, as well as on site temperatures, these three basic properties also determine the suitability of the site for insects. The general influence of site factors can be viewed as follows:

Specific topographic features (elevation, slope, and aspect)

PLUS

Specific soil properties (structure, chemistry, and bedrock formation)

PLUS

Specific weather conditions (precipitation, solar radiation, wind)

DETERMINE

General characteristics of the site, or habitat type (nutrients, moisture, and temperature)

WHICH INFLUENCE

Plant and animal reproduction, survival, and migration = population dynamics

WHICH DIRECT

Plant succession and the eventual climax association

Site factors may influence insect populations directly or indirectly. Temperature is probably the most important direct factor acting on cold-blooded insects, influencing their rates of reproduction, survival, and development, as well as their flight capacities (Chapter 1). Thus, warm, dry conditions, provided they are not too warm or dry, will tend to make insect environments more favorable. As we know, insect populations inhabiting favorable environments often attain higher equilibrium levels and may exhibit oscillations or cycles around equilibrium (Chapter 4). In general, then, insect populations are likely to be more dense, to exhibit greater fluctuations, and to cause more damage as site conditions become more xeric (drier and warmer).

Site factors may also affect insect reproduction and survival indirectly through their effect on the host plants. Sites that are deficient in certain nutrients or water, for example, may cause plants to grow under stressed conditions. This may adversely affect their defense systems or alter the nitrogen balance in their tissues and thus indirectly increase the reproduction or survival, or both, of insect populations (Chapters 3 and 4). As a general rule, therefore, insect populations are likely to be more dense, to exhibit greater fluctuations, and to cause more damage, when sites are deficient in nutrients or moisture. It is significant that, when the two site effects are taken together, site xericity has both direct and indirect effects on the abundance of insects, whereas nutrients only act indirectly through the host plant. This means that site xericity is likely to have a profound influence on insect population dynamics and should probably be included in most risk assessment models.

When considering site effects, we should realize that different tree species are adapted to particular site conditions. Some species, such as the ponderosa pine, are adapted to xeric sites, whereas others, such as the sitka spruce, require moist (mesic) sites. Thus, changes in site moisture must always be considered in relationship to the adaptations of the species. In other words, a dry site for sitka spruce will be a moist one for ponderosa pine. In nature, competition between tree species with different physiological requirements usually ensures that a particular site is occupied by species that are adapted to that site. Unadapted species will not be able to compete as well and will lose in the competitive struggle. Human interference with this natural process, however, sometimes results in tree species growing on sites to which they are poorly adapted. For instance, logging ponderosa pine and fire control practices in eastern Oregon and Washington have resulted in Douglas-fir growing on sites that were originally occupied by the pine. The growth of Douglas-fir on these drier than normal sites has led to some serious insect problems (see Chapter 11).

Our theoretical discussion indicates that site factors affecting soil moisture should receive special consideration when designing risk assessment models. Such variables as soil structure and depth, site elevation, slope and aspect, and precipitation and temperature may provide significant predictive capabilities.

TABLE 7.1
Site Variables Associated with the Probability of Insect Damage to Forests

Insect/tree	Variable	High risk	Sources
a. Mountain pine beetle/ lodgepole pine	Temperature	Hot summer	Safranyik *et al.* (1974)
	Precipitation	Low	Amman *et al.* (1977)
	Elevation	Low	
b. Southern pine beetle/ loblolly and shortleaf pines	Site index	High	Kushmaul *et al.* (1979)
	Soil pH	Low	
	Clay (%)	Low	Belanger *et al.* (1981)
	A horizon	Shallow	Lorio (1978)
	Moisture	Very high	
c. Fir engraver beetle/ grand fir	Understory plants	Dry sites	Schenk *et al.* (1976*a*)
d. Douglas-fir tussock moth/ grand fir and Douglas-fir	Slope	Ridgetops	Stoszek *et al.* (1981)
	Aspect	South facing	Heller and Sader (1980)
	Elevation	Low	
	Volcanic ash	Shallow	
e. European pine shoot moth/ red pine	Soil moisture	Low	Heikkenen (1981)
	Temperature	High	
f. Balsam woolly adelgid/ balsam and Fraser firs	Elevation	Low	Page (1975)
	Soil moisture	Low	
g. Pine leaf aphid/ pine and spruce	Soil structure	Sandy gravel	Dimond and Bishop (1968)
	Temperature	Cool	
h. Budworm/ balsam fir and spruce	Climate	Warm/dry	Blais and Archambault (1982)

These conclusions are supported by empirical studies demonstrating that insect damage is often associated with warm and dry conditions (Table 7.1). Exceptions, however, do occur (Table 7.1b,g). In the first case, southern pine beetle damage was found to be higher on very moist, waterlogged soils. This was probably because excessive soil moisture causes stress to the tree by preventing root aeration. In the second exception, the risk of aphid damage was higher on cooler sites because they supported the correct mixture of spruce and pine, both of which are required for the aphid to complete its life cycle. These exceptions to our general site rules illustrate the importance of careful study when designing risk assessment methods. Although the general rules may hold most of the time, they should not blind us to other important site factors that may predispose stands to insect attack under certain conditions.

7.1.2. Stand Factors

Although site factors set the stage on which the drama of stand development and insect outbreak takes place, the drama is also affected by interactions between the principal actors, the trees, and by the progress of time. For example, when stands become very dense and their crowns close, the photosynthesizing leaf area of each tree declines, making less carbohydrate energy available for dynamic defensive responses (Chapter 3). Thus, trees growing under competition stress are much more likely to succumb to insect attack. Age is another important factor in the ability of trees to defend themselves. As trees grow old, their physiological processes, including their defense reactions, become impaired by the aging process. As we can see from Table 7.2, stand density and age are often found to be important assessment variables.

The species composition of a stand may also directly affect the reproduction and survival of insect pests. For instance, insects feeding on a particular plant species will have a harder time finding their food if it is mixed among a lot of

TABLE 7.2
Stand Variables Associated with the Probability of Insect Damage to Forests

Insect/tree	Variable	High risk	Sources
a. Mountain pine beetle/	Age	Old	Safranyik *et al.* (1974)
lodgepole pine	Diameter	Large	Amman *et al.* (1977)
	Density	High	Schenk *et al.* (1980)
	Species	High % pine	
b. Mountain pine beetle/	Age structure	Even-aged	Sartwell and Stevens (1975)
ponderosa pine	Age	Old	
	Diameter	Medium	
	Density	High	
	Species	High % pine	
c. Southern pine beetle/	Age	Old	Lorio (1978)
loblolly and shortleaf pines	Disturbances	High	
	Density	High	Kushmaul *et al.* (1979)
	Species	High % pine	
	Understory	High cover	
d. Fir engraver beetle/	Density	High	Schenk *et al.* (1976*b*)
grand fir	Species	High % fir	
e. Spruce beetle/	Density	High	Schmid and Frye (1976)
Engelmann spruce	Species	High % spruce	
	Diameter	Large	
f. Budworm/	Age	Old	Morris and Bishop (1951)
balsam fir and spruce	Density	High	Blais and Archambault (1982)
	Species	High % fir	
g. Douglas-fir tussock moth/	Age	Old	Stoszek *et al.* (1981)
grand fir	Density	High	
	Species	High % fir	
	Crown levels	Multistoried	
h. Balsam woolly adelgid/	Density	High	Page (1975)
balsam and Fraser firs	Species	High % fir	
i. Pine leaf aphid/	Species	Pine and	Dimond and Bishop (1968)
pine and spruce		spruce equal	

nonhost trees. Similarly, if certain age classes of trees are susceptible or preferred, the survival of migrating insects will be lowered if the stand is composed of a mixture of age classes. Stand structure, in terms of species and age distributions, therefore, is often an important criterion for risk assessment (Table 7.2).

In addition to the direct effects, the dynamics of insect populations may be influenced indirectly through the effect of stand factors on natural enemy populations. For instance, woodpeckers and other hole-nesting insectivorous birds are dependent on dead trees for nesting sites, whereas insect predators and parasites often rest and feed on understory vegetation, using nectar from flowers as a source of energy. Although these indirect effects are often difficult to assess, they should at least be in our minds when we construct risk models.

7.1.3. Disturbing Factors

Certain environmental variables that change dramatically from time to time can either alter the average stand or site conditions or directly influence the insect population itself. Weather factors are particularly important in this respect, being notoriously changeable and unpredictable. Drought periods and warm, dry spring and summer temperatures are particularly significant in raising insect

TABLE 7.3
Disturbance Variables Associated with the Probability of Insect Damage to Forest Stands

Insect/tree	Variable	High risk	Sources
a. Western pine beetle/ ponderosa pine	Precipitation	Droughts	Miller and Keen (1960)
	Thunderstorms	Lightning strikes	
	Pathogens	Mistletoe and root diseases	
b. Fir engraver beetle/ grand and white firs	Precipitation	Droughts	Ferrell and Hall (1975)
	Insects	Defoliation	Wright et al. (1984)
	Pathogens	Root infections	Hertert et al. (1975)
	Temperature	Warm winters	Berryman (1970)
c. Spruce bark beetle/ Norway spruce	Wind	Many windthrows	Worrell (1983)
	Precipitation	Droughts	
d. Southern pine beetle/ loblolly and shortleaf pines	Thunderstorms	Lightning strikes	Lorio (1978)
	Wind	Windthrows	
	Precipitation	Flooding	
e. Douglas-fir beetle/ Douglas-fir	Pathogens	Mistletoe; root diseases	Furniss et al. (1981)
	Fire	Injured trees	
	Wind	Windthrows	
	Insects	Defoliation	Wright et al. (1984)
	Man	Logging	Lejeune et al. (1961)
f. European pine shoot moth/ red pine	Temperature	Warm winters	Heikkenen (1981)
	Precipitation	Summer drought	
g. Budworm/ balsam fir and spruce	Weather	Warm-dry spring	Greenbank (1956)

population levels and precipitating outbreaks (Table 7.3). By contrast, cool, moist summer temperatures and cold winters may drastically reduce insect population densities. Other weather disturbances, such as lightning and wind storms, as well as biological disturbances due to insects, pathogens, or man, may also result in increased damage to forest stands (Table 7.3).

7.1.4. Individual Tree Factors

Methods used to assess the probability of attack by insects on individual trees or on small groups of trees (e.g., bark beetles) are often based on individual tree characteristics. The two most commonly used variables are tree age and crown vigor (Fig. 7.1). However, some systems use a large number of tree characteristics (e.g., see Table 7.5).

Another approach to individual tree assessment is to use a variable that integrates all the effects of various factors acting on tree vigor. One of the most commonly used *integrative variables* is annual stem growth increment, or current radial increment (Fig. 7.2). It has been argued, however, that current growth increment may not be an accurate reflection of vigor because slow-growing trees may well be adapted to the particular site and stand conditions under which they are growing, hence are not necessarily under stress. This has prompted the

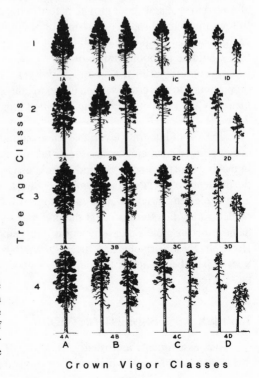

FIGURE 7.1. Keen's individual tree risk assessment method for ponderosa pine threatened by western pine beetle attack in the eastside Sierra Nevadas of California. This system has been modified in many ways over the years. (See Miller and Keen, 1960.)

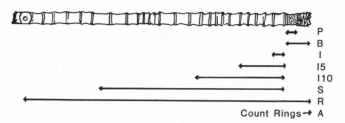

FIGURE 7.2. Some useful data obtained from an increment core: P, phloem thickness; B, bark thickness; I, current annual growth increment; I5, 5-year growth increment; I10, 10-year growth increment; S, sapwood depth; R, radius of tree; A, age of tree at the height at which core was extracted. The vigor of the tree can also be estimated as follows:

Periodic growth ratio (PGR) = I5/(I10–I5) (see Mahoney 1978)
Relative sapwood increment (RSI) = $[(R-B)^2 - (R-B-I)^2]/[(R-B)^2 - (R-B-S)^2]$
(see Waring and Pitman 1980)

hypothesis that vigor is related to the acceleration or deceleration of tree growth as reflected by the periodic growth ratio (PGR) (Fig. 7.2). Trees that show decelerating growth (PGR <1) are assumed to be under stress and susceptible to insect attack.

An alternative argument for an integrative variable is based on the energy budget of the tree. It is argued that energy, in the form of photosynthate, is allocated to root and shoot growth before being used for stemwood production. It follows that annual sapwood production should reflect the availability of surplus energy and that the proportion of the total energy allocated to stemwood production should be a good estimate of tree vigor. As the total energy produced by a tree is proportional to the area of photosynthesizing leaf tissue, which is itself proportional to the cross-sectional area of the sapwood, tree vigor can be estimated by the ratio of the current sapwood area to the total sapwood area (Fig. 7.2).

The use of integrative variables has some very desirable features. First, only one or a few measurements need to be taken, and this reduces the time and effort in making an assessment. Second, a truly accurate integrative variable measures the influence of all site, stand, and individual tree factors, while models based on a few selected variables may miss some of these. For example, assessment models rarely take into account the presence of root diseases or previous insect damage because their measurement is either impossible or too time consuming. However, these variables have been shown to be very important in predisposing trees to bark beetle attack (Table 7.3). A good integrative variable will estimate these effects.

7.2. RISK ASSESSMENT METHODS

Risk assessments are usually based on a formal examination of individual tree or stand conditions, which are then plugged into a predictive model. Various kinds of models will be described later in this section. Even if formal models are not available, however, assessments may still be made informally from a basic knowledge of insect population dynamics. For instance, we would expect more insect problems in dense stands with decelerating growth rates growing on dry sites. An informal risk assessment, based on general knowledge of insect–forest interactions, is far better than no assessment at all.

7.2.1. Classification Models

Classification systems are usually employed when quantitative data are unavailable or if the data are not amenable to statistical analysis. These models are frequently based on qualitative observation or personal experience concerning factors associated with insect outbreaks. For example, past experience and measurements in outbreak areas indicated that the probability of spruce beetle outbreaks was higher in dense, pure, large-diameter spruce stands growing on better spruce sites. On the basis of this information, a subjective classification system was developed for estimating the risk of beetle outbreaks in spruce stands (Table 7.4).

The major weakness of classification models is their imprecision and lack of statistical reliability. On the other hand, they are easy to use and also suggest practical solutions to the problem. We can see, for example, that the risk of spruce beetle damage can be reduced by controlling site, tree diameters, stocking density, or species composition.

TABLE 7.4
Criteria for Assessing the Risk of Spruce Beetle Outbreaks
in Engelmann Spruce Stands[a]

Risk[b] category	Physiographic location	Average diameter of live spruce > 10 in DBH (in DBH)	Basal area (ft²)	Proportion of spruce in canopy (%)
High (3)	Spruce on well-drained sites in creek bottoms	>16	>150	>65
Medium (2)	Spruce on sites with site index of 80 to 120	12–16	100–150	50–65
Low (1)	Spruce on sites with site index of 40 to 80	<12	<100	<50

[a]Modified from Schmid and Frye (1976).

[b]Risk is determined by assigning the number shown in parentheses under risk category to a stand under observation for each risk attribute and then summing them. For example, a stand growing on a well-drained creek bottom (3), with mean tree diameter of 14 in (2), basal area less than 100 (1), and more than 65% spruce (3), has the following risk value (RV): 3 + 2 + 1 + 3 = 9. Stands can then be assessed as low (RV = 4–6), moderate (RV = 7–9) or high risk (RV = 10–12).

7.2.2. Penalty-Point Models

A variation of the classification method is the so-called penalty-point system, in which penalties are assigned to trees or stands according to the presence of traits associated with risk. For example, in Table 7.5, penalties are assigned to specific tree characteristics considered important in determining the probability of its death, larger penalties being assessed to the more important variables. Penalty-point models are also easy to use but, like classification methods, are often statistically imprecise. Recently, however, quantitative techniques have become available for assigning penalties in a more objective fashion.

7.2.3. Mathematical Models

Mathematical techniques can be used for constructing assessment models, provided that data are available and are adequate to the task. Ideally, data should be obtained from a large number of stands (or individual trees) before the occurrence of outbreaks (or tree damage). Once these stands or trees have been selected, measurements are made of as many site, stand, or individual tree variables as possible, with particular attention being paid to those variables that the literature suggests are important in determining the degree of risk (Tables 7.1–7.3). The test stands or trees are then examined after a fixed time period, say 5–10 years, and the damage to each stand is estimated using standard methods (Chapter 6).

Although an *a priori* (before-the-fact) experimental design is the only foolproof way of eliminating sampling bias, it is sometimes impractical because of time constraints or the nonrandom distribution of insect infestations; for example, damage may be highly clumped, in which case outbreak stands will be undersampled. The alternative is a postmortem study, in which economically damaged trees or stands are selected after damage has occurred. Sampling bias can be reduced to some extent by selecting sampling units (stands or trees) at random within the general infested and uninfested areas (see Chapter 6 for sampling methods). We should realize, however, that in addition to sampling bias, some important variables may be difficult or impossible to measure in postmortem studies. For example, the phloem and sapwood may shrink considerably after tree death, and foliage characteristics are sometimes impossible to measure in heavy defoliated or dead trees.

7.2.3.1. Logistic Regression

The logistic equation is a useful model for describing the probability of discrete events as a sigmoid function of a set of independent site and stand variables (Table 7.6a). To use this method, the dependent variable must be

TABLE 7.5
Penalty System for Assessing the Risk of Western Pine Beetle Mortality to Individual Ponderosa and Jeffrey Pines in the Eastside Sierra Nevada Mountains[a]

Condition/factor	Penalty
Needle condition	
Needle complement	
Needle complement normal	0
Less than normal needle complement but no contrast between upper and lower crown	2
Thin complement in upper crown, normal in lower crown	5
Needle length	
Needle length normal	0
Needles shorter than normal throughout crown. No contrast between upper and lower crown	2
Needles short in top, normal below. Marked contrast	5
Needle color	
Normal	0
Off color	2
Fading	8
Twig and branch conditions	
No twigs or branches dead	0
A few scattered dead or dying twigs or branches in crown	1
Many scattered dead or dying twigs or branches in crown	2
Dead or dying twigs or branches in crown forming a definite weak spot in crown, notably in top one-third of crown	3
Dead or dying twigs or branches in crown forming more than one weak spot or hole, notably in top one-third of crown	5
Top crown conditions	
No top killing	0
Old top kill with no progressive weakness or killing in green crown below	5
Current top killing	8
Broken top—recent, less than one-third	5
Broken top—recent, more than one-third	8
Broken top—old, no progressive weakness	2
Other factors	
Lightning strikes—recently struck, no healing evident	8
Lightning strikes—healed strike	2
Dendroctonus valens attacks in base—current, successful	6
Dendroctonus valens attacks in base—old, pitched out	2

Risk class		Penalty score
Low	I	0
	II	1–4
	III	5–7
High	IV	≥8

[a]As reported by Smith *et al.* (1981).

TABLE 7.6
The Logistic and Discriminant Functions and Their Use in Risk Assessment

a. Logistic function
$$P = 1/[1 + \exp(-X')]$$
(Stage and Hamilton, 1981)

P = probability of outbreak or individual tree death

exp = base of natural logarithm

$X' = b_0 + b_1X_1 + b_2X_2 \cdots + b_nX_n$

X_i = value of a specific risk predictor, usually a stand, site, or individual tree variable selected by a screening procedure (Hamilton and Wendt, 1975)

b_i = parameters estimated by linear regression (Hamilton, 1974)

Example
$$X' = 9.643 + 0.032X_1 - 0.078X_2 - 0.047X_3 - 1.419X_4$$
(Ferrell, 1980; for white fir attacked by bark boring beetles)

X_1 = percentage of white fir live crown with upturned or horizontal branches

X_2 = code for white fir crown density; dense (0); ragged or one-sided (1); ragged, dead or flagged branches (2); both (1) and (2) = (3)

X_3 = percentage of crown ragged because of missing, dead or dying branches

X_4 = code for live bark visible on stem; live bark visible (1); live bark not visible (2)

P = probability of a white fir tree dying within 5 years

b. Discriminant function
$$Y = b_1X_1 + b_2X_2 + \cdots + b_nX_n$$

Y = discriminant score

X_i = value of a specific risk predictor, usually a stand, site, or individual tree variable

b_i = parameters estimated by discriminant analysis

Example:
$$Y = -1.5X_1 + 0.93X_2 + 3.3X_3 + 64.3X_4$$
(Ku *et al.*, 1981, for southern pine beetle infestations in Arkansas)

X_1 = total basal area of the stand/acre

X_2 = basal area of hardwoods

X_3 = average stand age

X_4 = average 10-year radial increment in inches

Risk categories

High risk	$Y < 1$
Moderate	$1 < Y < 100$
Low risk	$Y > 100$

discrete so that the data can be separated into distinct classes, such as live and dead trees and outbreak and nonoutbreak stands. Data-screening procedures are then employed to determine the significant independent variables to be included in the model, and the parameters of the model are estimated by multiple regression (Table 7.6a).

7.2.3.2. Discriminant Analysis

The discriminant analysis method requires that the data be segregated into discrete classes. The analysis then identifies the set of independent site and stand variables that best discriminate between the dependent variable classes, that is, between live and dead trees, damaged and undamaged stands, and so forth. The analysis usually proceeds in a stepwise fashion, with variables added in order of significance toward enhancing the discrimination. Thus, discriminant analysis screens the variables for their relative significance and then produces a discriminant function out of the most significant or useful independent variable set (Table 7.6b).

As compared with the logistic equation, discriminant analysis has both advantages and disadvantages. If the statistical assumptions underlying the analysis are met, discriminant analysis may be the more effective method. However, the assumptions that the samples are drawn from multivariate normal populations with equal covariance matrices can seriously constrain the analysis. If these assumptions cannot be met, the logistic function may be a more appropriate model. In addition, the logistic function predicts the *probability* of economic damage occurring in contrast to the rather arbitrary risk categories assigned to discriminant functions. This direct probability assessment will often be of more value to forest managers.

7.2.3.3. Linear Regression

Linear regression is an appropriate model when the dependent variable is a continuous function of the independent variables, for example, if the number of trees expected to die from insect attack is expressed as a continuous linear function of site and stand variables. Even if the dependent variable is not linearly related to the independent variables, the data can often be transformed to an approximately linear form.

Multiple regression analysis usually proceeds in a stepwise fashion, with independent variables added to the model in order of predictive significance. The final model may consist of a single simple or complex variable or a string of variables (Table 7.7). The predictive power of the model is given by the R^2 value; i.e., $0 \leq R^2 \leq 1$ = the proportion of the total variation in the data explained by the model.

The main problem with mathematical equations is that complex computations may have to be carried out, and it is not always clear to the user how the variables can be manipulated in order to minimize the danger of insect-caused damage. The first problem can be overcome by the use of programmable pocket calculators. The second, however, is not so easily solved. One approach is to use the mathematical model to design a classification system (see Table 7.4) or a penalty system (such as that shown in Table 7.5). In this way the penalty points

TABLE 7.7
Regression Models for Assessing Damage Due to Forest Insect Infestations

Regression Equation

$$Y = a + b_1X_1 + b_2X_2 + \cdots + b_nX_n$$

Y = expected level of damage

a = regression intercept

X_i = value of a specific damage predictor variable, usually a stand or site variable

b_i = specific parameter value estimated by regression analysis

Example 1

$$Y = -0.681 + 0.447X_1 - 0.012X_2 + 0.505X_3 + 0.487X_4 + 0.274X_5, \ R^2 = 0.52$$

(Stozek *et al.*, 1981)

Y = $\log_e$ (percentage defoliation by the Douglas-fir tussock moth)

X_1 = topographic position (0 = lower slope, 1 = upper slope)

X_2 = depth of volcanic ash deposit in meters

X_3 = $\log_e$ (age of host trees at breast height)

X_4 = $\log_e$ [total stand basal area (m²/ha)/Douglas-fir site index], where site index = height (meters) at 50 yr

X_5 = $\log_e$ (percentage of stand basal area in grand fir)

Example 2

$$Y = -48.07 + 19.36 \exp(X_1 \cdot X_2), \ R^2 = 0.89$$

(Schenk *et al.*, 1980)

Y = percentage of stand basal area killed by the mountain pine beetle

X_1 = crown competition factor

X_2 = proportion of stand basal area in lodgepole pine

$\exp$ = base of the natural logarithm

In this example, the single predictor is a compound variable (the product of X_1 and X_2) that is nonlinearly (exponentially) related to Y.

---

or class boundaries are assigned specific probabilities by the mathematical equation. A combination of classic classification approaches and quantitative analysis may lead to assessment systems that possess the advantages of both approaches.

7.2.4. Graphical Models

In the preceding discussion we examined assessment models ranging from subjective classification systems to quantitative mathematical equations. All these methods have a common objective—that of identifying those variables that are most useful for assessing outbreak risk from a large number of site and stand variables. These methods are generally satisfactory for predicting gradient outbreaks, which are largely caused by the average characteristics of site and stand (Chapter 4). We should realize, however, that physical or biotic disturbances (Table 7.3) can drastically alter the risk of eruptive outbreaks, as they can be triggered by temporary environmental variations. In addition, the risk of eruptions is affected by the proximity of stands to other infested stands, because migrating beetles can initiate the explosion (Chapter 4). For these reasons, more complex models are sometimes required for predicting eruptive pest outbreaks.

One approach is to monitor the disturbance variables continuously and then update the general assessment according to current conditions. This approach is used for predicting fire danger because ignition risks change with current weather conditions. Similar procedures could be developed for assessing the danger of insect pest eruptions.

In Chapter 4 we saw that pest eruptions only occur when populations exceed a critical threshold density: the outbreak threshold. We also saw that thresholds are very sensitive to environmental conditions (see Fig. 4.4d); i.e., they will change in response to variations in site, stand, and weather conditions. In the case of bark beetles, for example, outbreak thresholds change as a direct function of stand vigor, or resistance to attack. This means that outbreaks can be initiated in very weak stands by a few beetles but can only be triggered in healthy (resistant) stands by very large beetle populations. This idea is diagrammatically represented in Figure 7.3. Here the environmental variables determining the outbreak threshold are interpreted more broadly because thresholds can be affected by factors other than stand resistance. For instance, the outbreak thresholds for gypsy moth and spruce budworm populations seem to be related to the density of vertebrate predators (see Chapter 5).

If we examine Figure 7.3, we see that outbreaks are initiated whenever the system enters the zone to the left of the threshold. This may occur if the regulating processes weaken (e.g., drought, defoliation, or disease lowering stand resistance from X to B) or if populations increase sufficiently (e.g., beetles breeding in windthrows, logging debris, or immigrating from adjacent stands

FIGURE 7.3. Relationship between the outbreak population threshold which separates low-density (endemic) from outbreak (epidemic) population dynamics and the strength of the population regulating factors. A system at position X can be moved into the outbreak zone if the population increases (X → A) or if the regulating process weakens (X → B): Once the system has moved into the outbreak zone (X → B), so that the insect population grows to C, then a strengthening of the regulating process may fail to terminate the outbreak (C → A). The system at position Y has a lower outbreak risk because larger disturbances, which occur less frequently, are required to push it over the outbreak threshold.

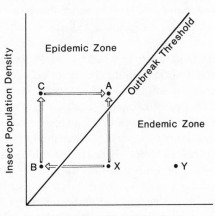

raising the population from X to A). In Figure 7.3 we can also see how temporary stress can precipitate large-scale epidemics. For instance, if a drought were to lower stand resistance from X to B, allowing the insect population to increase to C in the weakened trees, the outbreak would continue even after stand resistance would have returned to normal (C to A) because the insect population is still above the outbreak threshold.

It is also possible to see how the concept of risk can be introduced into threshold models. For example, if we have a stand at position X, then either a disturbance of magnitude X–B to the regulating process or one of magnitude A–X to the insect population will be sufficient to trigger an outbreak. In a stand with stronger regulating factors (e.g., a more resistant stand at position Y), however, the system would not enter the outbreak zone after disturbances of similar magnitude. Because small environmental disturbances occur more frequently than large ones, the stand at position X is at greater risk than that at Y. This theoretical result provides a basis for designing risk assessment models for eruptive pest populations.

In order to construct threshold risk models we have to be able to measure two critical variables: (1) the strength of the regulating factors, and (2) the density of the pest population (Fig. 7.3). We also need to determine the location of the outbreak threshold in relationship to these variables.

Let us start with the easiest problem, the estimation of pest population density. The most obvious approach is to sample either the insects themselves or their damage (see Chapter 6 for examples of sampling techniques). However, these procedures are usually labor intensive and expensive, particularly if population estimates have to be updated annually. Another possibility is to estimate the potential insect population from stand or site factors, or both, associated with insect abundance. Endemic bark beetle population densities, for example, are related to the number of trees weakened by lightning strikes, root diseases, windthrow, and the like and to the quantity and quality of phloem or bark in these trees (i.e., the quantity and quality of the food supply).

The next problem is to estimate the strength of the regulating factors as a function of site and stand variables or an integrating variable. For example, if host resistance limits the insect population, then we can express its strength in terms of soil depth and moisture, stand age and density, and so forth. Alternatively, as resistance is related to stand vigor, an integrative variable, such as that obtained by boring a sample of trees, could be used (Fig. 7.2).

Finally, we must locate the outbreak threshold as a function of the two critical variables. Because thresholds are transient states, they cannot be observed or estimated directly. Nevertheless, they can be approximated if the critical variables are measured in a large number of separate stands, some of which are undergoing outbreaks while others have low insect populations. If these data are then plotted, the threshold will appear as an imaginary line segregating the two groups of stands (Fig. 7.4). Aberrant data points will frequently

FIGURE 7.4. Threshold model for assessing the risk of mountain pine beetle outbreaks in lodgepole pine stands. (After Berryman, 1980, 1982.) Expected beetle population size is assumed to be directly related to the number of thick-phloemed trees in the stand (Amman *et al.*, 1977), whereas the strength of the population regulating factors is assumed to be directly related to stand resistance as estimated by periodic growth ratio (PGR), crown competition × proportion of stand basal area in lodgepole pine (SHR), and average stand age (Mahoney, 1978). Outbreak risk is assumed to be inversely related to the distance from the outbreak threshold. The approximate location of the threshold was found by measuring phloem thickness and stand resistance in a series of outbreak (●) and nonoutbreak (○) stands.

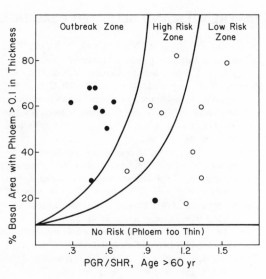

be observed because of inaccuracies in estimating the critical variables. When there is considerable overlap around the threshold, regression techniques can be used to determine the best estimate of the threshold and its confidence limits.

Once the threshold function has been found, the next step is to develop a risk decision plan. Previously we observed that the likelihood of an outbreak erupting in a particular stand depends on the distance between the outbreak threshold and the estimated position of the stand. In Figure 7.3, for instance, the stand at position X is at greater risk than that at Y because it can be moved into the outbreak domain by smaller disturbances. On this basis, risk decision lines can be assigned subjectively to produce a risk assessment model (Fig. 7.4). Managers will usually decide on the width of the decision zones on the basis of their own risk aversion, with some preferring to "play it safe," assigning wider risk zones than would be assigned by those who are willing, or are forced, to take risks.

Even though outbreak thresholds are difficult to estimate, once defined they provide a useful tool for predicting the likelihood of eruptive pest outbreaks. Without such methods, the manager is in constant danger of facing pest eruptions that can spread rapidly over huge forested areas. Experience has shown that once eruptive outbreaks have started they are extremely difficult to stop. It is therefore imperative that the manager identify and treat the high-risk stands, and this can only be done effectively with risk assessment models.

REFERENCES AND SELECTED READINGS

Amman, G. D., McGregor, M. D., Cahill, D. B., and Klein, W. H., 1977, Guidelines for reducing losses of lodgepole pine to the mountain pine beetle in unmanaged stands in the Rocky Mountains, U.S. Forest Service, General Technical Report INT-36, 19 pp. (risk classification)

Belanger, R. P., Porterfield, R. L., and Rowell, C. E., 1981, Development and validation of systems for rating the susceptibility of natural stands in the Piedmont of Georgia to attack by the southern pine beetle, in: Hazard-rating systems in forest pest management (R. L. Hedden, S. J. Barras, and J. E. Coster, eds.), U.S. Forest Service, General Technical Report WO-27, pp. 79–86. (discriminant analysis)

Berryman, A. A., 1970, Overwintering populations of *Scolytus ventralis* (Coleoptera:Scolytidae) reduced by extreme cold temperatures, *Ann. Entomol. Soc. Am.* **63:**1194–1196.

Berryman, A. A., 1980, General constructs for risk decision models, *Proceedings of the Society of American Foresters, Spokane,* pp. 123–128. (threshold risk models)

Berryman, A. A., 1981, Effects of site characteristics on insect population dynamics, *Proc. XVII IUFRO World Congr. Kyoto* **2:**541–549.

Berryman, A. A., 1982, Biological control, thresholds, and pest outbreaks. *Environ. Entomol.* **11:**544–549. (threshold concepts)

Berryman, A. A., and Stark, R. W., 1985, Assessing the risk of forest insect outbreaks, *Z. Angew. Entomol.* **99:**199–208.

Bevan, D., and Stoakley, J. T. (eds.), 1985, Site Characteristics and Population Dynamics of Lepidopteran and Hymenopteran Forest Pests, Forestry Commission Research and Development Paper 135, Edinburgh.

Blais, J. R., and Archambault, L., 1982, Rating vulnerability of balsam fir to spruce budworm attack in Quebec, Canadian Forest Service, Information Report LAU-X-51, 19 pp.

Dimond, J. B., and Bishop, R. H., 1968, Susceptibility and vulnerability of forests to the pine leaf aphid, *Pineus pinifoliae* (Fitch) (Adelgidae), Maine Agricultural Experiment Station, Bulletin 658, Orono, 16 pp. (regression analysis)

Ferrell, G. T., 1980, Risk-rating systems for mature red fir and white fir in Northern California, U.S. Forest Service, General Technical Report PSW-39, 19 pp. (penalty-point system)

Ferrell, G. T., and Hall, R. C., 1975, Weather and tree growth associated with white fir mortality caused by fir engraver and roundheaded fir borer, U.S. Forest Service, Research Paper PSW-109, 11 pp.

Furniss, M. M., Livingston, R. L., and McGregor, M. D., 1981, Development of a stand susceptibility classification for Douglas-fir beetle, in: Hazard-Rating Systems in Forest Pest Management (R. L. Hedden, S. J. Barras, and J. E. Coster, eds.), U.S. Forest Service, General Technical Report WO-27, pp. 115–128.

Greenbank, D. O., 1956, The role of climate and dispersal in the initiation of outbreaks of the spruce budworm in New Brunswick. I. The role of climate, *Can. J. Zool.* **34:**453–476.

Hamilton, D. A., Jr., 1974, Event probabilities estimated by regression, U.S. Forest Service, Research Paper INT-152, 18 pp. (fitting logistic function)

Hamilton, D. A., Jr., and Wendt, D. L. R., 1975, SCREEN: A computer program to identify predictors of dichotomous dependent variables, U.S. Forest Service, General Technical Report INT-22, 19 pp. (screening out significant variables)

Heikkenen, H. J., 1981, The influence of red pine site quality on damage by the European pine shoot moth, in: Hazard-rating systems in forest insect pest management (R. L. Hedden, S. J. Barras, and J. E. Coster, eds.), U.S. Forest Service, General Technical Report WO-27, pp. 35–44. (regression analysis)

Heller, R. C., and Sader, S. A., 1980, Rating the risk of tussock moth defoliation using aerial photographs, U.S. Department of Agriculture Handbook 569, 21 pp.

Hertert, H. D., Miller, D. L., and Partridge, A. D., 1975, Interaction of bark beetles (Coleoptera: Scolytidae) and root-rot pathogens in grand fir in northern Idaho, *Can. Entomol.* **107:**899–904.

Ku, T. T., Sweeney, J. M., and Shelburne, V. B., 1981, Hazard rating of stands for southern pine beetle attack in Arkansas, in: Hazard-rating systems in forest insect pest management (R. L. Hedden, S. J. Barras, and J. E. Coster, eds.), U.S. Forest Service, General Technical Report WO-27, pp. 145–148. (discriminant analysis)

Kushmaul, R. J., Cain, M. D., Rowell, C. E., and Porterfield, R. L., 1979, Stand and site conditions related to southern pine beetle susceptibility, *For. Sci.* **25:**656–664. (discriminant analysis)

Lejeune, R. R., McMullen, L. M., and Atkins, M. D., 1961, The influence of logging on Douglas-fir beetle populations, *For. Chron.* **37:**308–314.

Lorio, P. L., Jr., 1978, Developing stand risk classes for the southern pine beetle, U.S. Forest Service, Research Paper SO-144, 9 pp.

Lorio, P. L., Jr., Mason, G. N., and Autry, G. L., 1982, Stand risk rating for the southern pine beetle: Integrating pest management with forest management. *J. For.* **80:**212–214. (discriminant analysis)

Mahoney, R. L., 1978, Lodgepole pine/mountain pine beetle risk classification methods and their application, in: Theory and Practice of Mountain Pine Beetle Management in Lodgepole Pine Forests (A. A. Berryman, G. D. Amman, R. W. Stark, and D. L. Kibbee, eds.), pp. 106–113, Forest Wildlife and Range Experiment Station, University of Idaho, Moscow, 224 pp. (risk classifications, regression analysis, threshold models)

Miller, J. M., and Keen, F. P., 1960, Biology and control of the western pine beetle, U.S. Forest Service, Miscellaneous Publication No. 800, 381 pp. (risk classifications and penalty point systems)

Morris, R. F., and Bishop, R. L., 1951, A method of rapid forest survey for mapping vulnerability to spruce budworm damage, *For. Chron.* **27:**1–8. (risk classification)

Page, G., 1975, The impact of balsam woolly aphid damage on balsam fir stands in Newfoundland, *Can. J. For. Res.* **5:**195–209.

Safranyik, L., Shrimpton, D. M., and Whitney, H. S., 1974, Management of lodgepole pine to reduce losses from the mountain pine beetle, Environment Canada, Forest Service, Technical Report 1, 24 pp.

Sartwell, C., and Stevens, R. E., 1975, Mountain pine beetle in ponderosa pine: Prospects for silvicultural control in second-growth stands, *J. For.* **73:**136–140. (risk variables)

Schenk, J. A., Mahoney, R. L., Moore, J. A., and Adams, D. L., 1976*a*, Understory plants as indicators of grand fir mortality due to the fir engraver, *J. Entomol. Soc. BC* **73:**21–24. (regression analysis)

Schenk, J. A., Moore, J. A., Adams, D. L., and Mahoney, R. L., 1976*b*, A preliminary hazard rating of grand fir stands for mortality by the fir engraver, *For. Sci.* **23:**103–110. (regression analysis)

Schenk, J. A., Mahoney, R. L., Moore, J. A., and Adams, D. L., 1980, A model for hazard rating lodgepole stands for mortality by mountain pine beetle, *For. Ecol. Manage.* **3:**57–68. (regression analysis)

Schmid, J. M., and Frye, R. H., 1976, Stand ratings for spruce beetle, USDA Forest Service, Research Note RM-309, 4 pp. (risk classification)

Smith, R. H., Wickman, B. E., Hall, R. C., DeMars, C. J., and Ferrell, G. T., 1981, The California pine risk-rating system: Its development, use, and relationship to other systems, in: Hazard-rating systems in forest insect pest management (R. L. Hedden, S. J. Barras, and J. E. Coster, eds.), U.S. Forest Service, General Technical Report WO-27, pp. 53–69. (risk classification and penalty point systems)

Stage, A. R., and Hamilton, Jr., D. A., 1981, Sampling and analytical methods for developing risk rating systems for forest pests, in: Hazard-rating systems in forest insect pest management (R. L. Hedden, S. J. Barras, and J. E. Coster, eds.), U.S. Forest Service, General Technical Report WO-27, pp. 87–92. (logistic function)

Stoszek, K. J., Mika, P. G., Moore, J. A., and Osborne, H. L., 1981, Relationships of Douglas-fir tussock moth defoliation to site and stand characteristics in Northern Idaho, *For. Sci.* **27**:431–442. (regression analysis)

Waring, R. H., and Pitman, G. B., 1980, A simple model of host resistance to bark beetles, Oregon State University, School of Forestry (Corvallis), Forest Research Laboratory, Research Note 65, 2 pp. (threshold model)

Worrell, R., 1983, Damage by the spruce bark beetle in South Norway 1970–80: A survey, and factors affecting its occurrence, Reports of the Norwegian Forest Research Institute 38.6, 34 pp.

Wright, L. C., Berryman, A. A., and Wickman, B. E., 1984, Abundance of fir engraver, *Scolytus ventralis,* and the Douglas-fir beetle, *Dendroctonus pseudotsugae,* following tree defoliation by the Douglas-fir tussock moth, *Orgyia pseudotsugata, Can. Entomol.* **116**:293–304.

CHAPTER 8

PREVENTION OF INSECT OUTBREAKS

In the professional disciplines of forestry, agriculture, and medicine, the practice of prevention involves the manipulation of environmental conditions so that they become unfavorable for the reproduction and survival of pest organisms. In this way pest populations are kept at low densities and outbreaks are avoided. For example, medical practitioners may prescribe a change of diet or life-style designed to reduce stress and thereby improve the resistance of their patients to disease-causing microorganisms. What they are doing, in fact, is making the environment unfavorable for the microorganism by "tuning up" the patient's defense systems. Similarly, the forester may prescribe fertilization or thinnings to reduce the stress on forest stands and so improve their defenses against insect attack. Because preventive treatments rely on an understanding of the relationships between the pest organism and its environment, prevention is really the *practice of applied ecology.*

In Part II we showed that populations can only be regulated at low densities by the action of negative feedback processes and that negative feedback often results from trophic, or feeding, interactions. These general principles lead to the conclusion that pest populations can be regulated at low densities by interactions with their food supply, in our case forest trees, or by interactions with their predators, parasites, or pathogens (Fig. 8.1). The manager can enter this system by manipulating the plant, natural enemy, or pest populations in such a way that the pressure on the pest is increased. Manipulation of the tree population, or control "from below," is usually referred to as *silvicultural control,* whereas manipulation of natural enemy populations, or control "from above," is often called *biological control.* These terms are somewhat misleading, however, because both approaches involve biological manipulations, and natural enemy populations can obviously be affected by silvicultural practices.

145

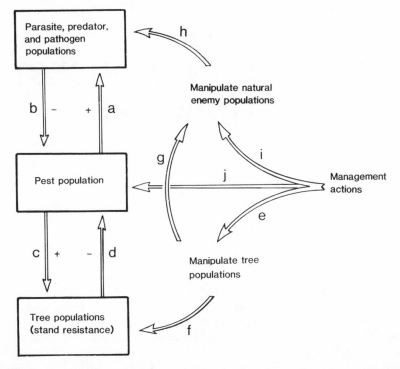

FIGURE 8.1. Feedback diagram illustrating pest population regulation by negative feedback. Pest populations can be regulated by natural enemies "from above," that is, increases in pest density induce changes in natural enemy populations (a), which then negatively impact pest population density (b), or by plant population "from below"; i.e., increases in pest density induce changes in plant resistance (c), which reduces pest density (d). Managers can manipulate this feedback system by altering plant populations to increase defensive capabilities (e,f) or, indirectly, natural enemy effectiveness (e,g,h), or by directly altering natural enemy effectiveness (i,h). Although a more risky proposition, managers can also manipulate the pest population to increase natural enemy effectiveness (j,a) or plant resistance (j,c). (Modified from Berryman, 1982.)

In this chapter, we will examine the effects of silvicultural practices on populations of insect pests and their natural enemies and try to see how these practices can be used to minimize the likelihood of destructive outbreaks. We will also discuss the possibilities of preventing pest outbreaks by manipulating natural enemy populations or, in certain cases, by manipulating the pest population itself.

8.1. SILVICULTURAL PRACTICES

Silviculture is the art of growing forests to certain predefined specifications. It is an art because it requires human imagination, innovation, and creativity and, like all creative endeavors, uses scientific principles and technology to achieve the desired effect—if the canvas of the silviculturist is the forest, the colors are ecological principles and the brush is the chainsaw. No matter how well the brush is wielded, an accurate picture cannot be created without a fundamental understanding of evolution, population dynamics, competition, and predation. The silviculturist must therefore also be schooled in the fundamental theories of plant and animal interactions; i.e., the science of ecology.

The practicing silviculturist uses ecological and technical knowledge to create forest structures designed by management land-use planners. The design may be very intricate, containing populations of fish, game animals, birds, cattle, and humans, as well as trees of different species and ages. With this conception in mind, the silviculturist *diagnoses* the present condition of the forest to ascertain how well it conforms to the desired abstraction and then *prescribes* various treatments that, it is hoped, will lead to the development of the planned effect. The kinds of prescriptive treatments available include harvest or preharvest timber cuts; site improvements such as burning, fertilization, and scarification; and regeneration procedures such as planting, genetic improvement, brush control, and so on. These treatments will have a decisive influence on the structure of the future forest, in particular the mixture of species present, the age and size distributions of the trees, and stand density. They will also influence populations of herbivorous insects and their natural enemies, causing pest populations to either increase or decrease. In the following sections we will examine some of the effects of silvicultural treatments on insect populations and try to see how these activities can be employed to prevent destructive outbreaks.

8.1.1. Cutting Methods

A number of cutting methods can be employed for harvesting and regenerating forest stands, or for controlling stand density and species composition. When employing these methods the silviculturist should be concerned with their effects on future insect populations.

8.1.1.1. Clearcutting

As the name implies, clearcutting involves the removal of all or most of the trees (some trees may be left to produce seed) over a relatively large area of land. Clearcutting is often considered a cost-effective harvesting method because only one cut is required and heavy machinery can be used for skidding and yarding the logs. Clearcutting, however, produces large quantities of logging debris (slash), may cause soil erosion and compaction, and may require seed-bed preparation (burning or soil scarification) and planting, all of which increase the cost of operation.

Clearcutting may also give rise to severe insect problems. For example, many species of *Ips* beetles and some *Dendroctonus* beetles (e.g., the European spruce beetle, *Ips typographus*, and the Douglas-fir beetle, *Dendroctonus pseudotsugae*) breed in accumulations of logging slash and may emerge later to attack and kill trees bordering the clearcut. In areas in which these insects are problems, slash should be disposed of by burning, removing the bark, or treatment with insecticides. The problem of insect infestations in slash can also be dealt with by planning clearcuts so that insect populations move from one slash pile to another. In the Douglas-fir region of the Pacific Northwest, for example, clearcuts are planned so that they are close enough to those of the previous year for Douglas-fir beetle populations to move from one slash pile to the next rather than into standing trees.

Reproduction on clearcuts can also be severely damaged by insect infestations. The use of heavy machinery frequently causes soil compaction, particularly in clay/loam soils, so that young seedlings have problems establishing root systems. In this weakened state, they often become vulnerable to attack by shoot and root borers. Tip and shoot insects, in particular, seem to prefer open, sunny, environments and often cause heavy damage to reproduction on clearcuts. Other insects such as *Hylobius* weevils breed in the many tree stumps and may then seriously damage or kill seedlings by feeding on the stems and roots.

The edges of clearcuts are particularly vulnerable to insect attack. Edge trees may suffer from shock when their crowns are suddenly exposed to full sunlight, or their roots and stems may be damaged by heavy machinery or falling trees. The weakened trees may then be attacked by beetles and other stem borers. The edges of clearcuts are also susceptible to windthrow and, should this occur, bark beetle populations often build up in the fallen timber and later attack and kill adjacent standing trees. Nowadays, most clearcuts are planned to minimize the impact of logging and environmental factors on the edge trees by orientating the cuts along natural discontinuities or topographies, such as streams, stand boundaries, and wind flow contours.

To ensure that future stands will grow under the most favorable conditions, hence be more resistant to insect attack, it is important that the species to be regenerated and their genetic stock be carefully matched to the prevailing site conditions. If decisions concerning stand regeneration are based on economic rather than ecological reasoning, they may lead to insect-susceptible forests because the trees are not genetically adapted to the particular site conditions. Natural selection is probably the best indicator of which tree species and strains are well adapted to the site. Thus, the selection of species for planting should reflect the natural forest composition. This can usually be determined by referring to the original timber type maps of the virgin forest.

The genetic composition (species and strain) of regeneration on clearcut areas can be controlled by leaving selected trees standing to provide a seed source or by planting nursery stock. Ideally, both methods should select trees for their insect and disease resistance, as well as their growth potential. If insect and disease resistance cannot be measured, the most vigorous individuals should be left as seed trees, and planting stock should be matched to the correct site conditions (i.e., the parent trees from which the seedlings came should be growing in a similar climatic regime, or provenance).

Artificial regeneration provides the silviculturist with the opportunity to plant genetically superior trees. However, genetic improvement programs are often designed to improve growth and yield rather than insect resistance. The genes control both the efficiency of the tree in converting sunlight into carbohydrates and its priority for using the energy for growth, reproduction, and defense. Breeding trees for fast growth may therefore produce strains that allocate little energy to defense. Unless extreme care is practiced, forestry can easily fall into the same trap as agriculture where inbreeding for increased yield has resulted in crops that are highly susceptible to insects and diseases. Unlike agriculture, however, the continuous spraying of susceptible crops with pesticides is rarely profitable in forestry. It should also be noted that trees have slow turnover rates relative to their insect pests. Thus, even if resistant strains are produced, we should expect insects to adapt to these strains within the life-span of the tree.

Clearcutting regenerates stands with a more or less even age distribution. In fact, the utilization of clearcutting methods is part of what is known as *evenaged management*. Once established, an evenaged forest has a rather uniform, single-storied canopy, and the trees have a narrow diameter and height distribution (Fig. 8.2a). These characteristics are often desirable in commercial forests because log sizes are fairly uniform and fewer entries are needed for density control and harvest. An evenaged structure can pose some serious insect problems, however.

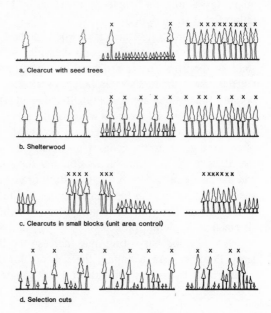

a. Clearcut with seed trees

b. Shelterwood

c. Clearcuts in small blocks (unit area control)

FIGURE 8.2. Effects of various cutting methods on the age structure of forests. x = trees selected for the next harvest cut.

d. Selection cuts

For instance, if certain age classes are more susceptible to insect attack, then all the trees in the stand will reach the susceptible age simultaneously. This problem can be partly solved by controlling the length of the rotation period, making sure that the stands are harvested before they reach susceptible ages. Evenaged stands, however, can also be weakened by other stress factors, such as drought or overstocking. The risk of extensive losses after events that stress younger stands can be reduced by encouraging diversity in species composition, and by making smaller clearcuts so that the forest as a whole has a more diverse age structure; i.e., during insect outbreaks, large pure stands are obviously going to experience greater losses than small mixed-species stands.

Long-term planning for clearcutting on a large forest involves many ecological, economic, and social considerations. In regions in which insect problems are particularly severe, entomological considerations may also be important in the planning process. It is here that risk assessment methods become invaluable to the silviculturist and forest manager, helping them to schedule clearcuts before stands become susceptible to insect attack (Chapter 7).

8.1.1.2. Shelterwood Harvesting

This method is usually employed when the manager wants to regenerate tree species with unpredictable or sparse seed crops, or when seedlings and saplings require protection from extremes of light, heat, wind, or frost. Unlike clearcutting, where only one harvest cut is made, the shelterwood method requires two or more entries into the stand, and this may increase harvesting costs. These economic losses may be balanced, however, by lower regeneration costs on naturally seeded shelterwoods.

Like clearcutting, the shelterwood method gives rise to stands containing trees of more or less equal age and size (Fig. 8.2b). These evenaged stands pose some of the same insect problems as clearcutting. For example, care must be taken to match the tree species being regenerated to the particular site conditions because species that are not optimally adapted may become stressed later in life. When this occurs, all the trees will tend to enter the insect-susceptible stage at the same time and extensive damage may occur. Shelterwood harvests, however, are often more gentle to the ecology than clearcuts. In particular, exposure shock at the edges of the cut and soil erosion and compaction are usually not as severe, and the shade provided by the overstory often reduces the incidence of tip and shoot insect damage in the reproduction. Shelterwood cutting also favors more shade-tolerant later successional species, which are sometimes more insect-resistant than intolerant pioneers.

Shelterwood harvesting may increase the likelihood of other insect problems, however. For example, the risk of blowdowns is often quite high, and windthrown trees may provide a food source for bark beetle populations.

Damage to standing trees and to reproduction during felling and skidding may also be high, and damaged trees may become susceptible to insect attack.

8.1.1.3. Unit Area Control

This harvesting method involves clearcutting and regenerating small "unit areas" of forest (6 to 10 acres) as evenaged blocks (Fig. 8.2c). This method increases the diversity of age structure over the entire forest by creating a mosaic of small evenaged stands. As such, it probably decreases the likelihood of extensive insect damage because stands of susceptible age are mixed among resistant stands. Otherwise, harvesting by unit area control poses much the same kinds of insect problems as clearcutting but on a smaller scale. Edge effects may be particularly troublesome, however, because smaller clearcuts cause a substantial increase in the total length of exposed stand edges in the forest; i.e., the ratio of edge to area increases as the clearcut gets smaller.

8.1.1.4. Selection Cutting

In this approach, individual trees or small groups of trees are harvested according to their own special characteristics, such as size, age, species, or risk of attack by insects and pathogens. The usual procedure is to harvest the larger mature and overmature trees, as well as the smaller high-risk trees, at regular intervals, say every 20 years. Removal of individual trees opens up the stand, releases suppressed understory trees, and provides small openings in which reproduction can become established. Continuous selection cutting results in an unevenaged stand with a multistoried canopy and a wide distribution of diameter classes (Fig. 8.2d). Because of this, selection is the basic cutting method used in *unevenaged management.*

The ability to select trees for cutting based on their individual merits gives the forester a high degree of control over stand structure and composition. Obviously, species composition can be altered by harvesting certain species more heavily than others. As with evenaged management, however, care must be taken to encourage those species which are best adapted to a particular site. In the past, where selection was based on economic rather than ecological reasons (i.e., "high-grading"), insect problems were often aggravated. Unevenaged management will favor more shade-tolerant species. Nevertheless, intolerant pioneer species can be encouraged by cutting groups of trees to allow more light to penetrate to the soil surface, but at some point *group selection* cutting merges into small clearcuts or unit area control. In general, selection cutting is more suitable for regenerating tolerant species, whereas clearcutting and its modifications are usually better suited to intolerant species. There are exceptions, however, such as the Eastside Sierra Nevada ponderosa pine type, where intolerant pines grow in pure stands with unevenaged structure. Selection cutting under

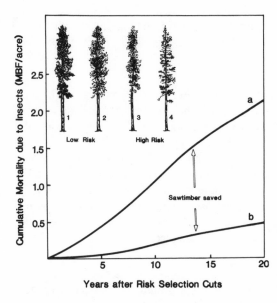

FIGURE 8.3. Cumulative ponderosa pine mortality caused by (a) insects in uncut compartments, and (b) compartments selectively logged for high-risk trees. All class 3 and 4 trees were cut, together with some lower risk trees to make up a minimal harvest of 2.5 MBF (= thousand board feet)/acre (Blacks Mountain Experimental Forest, California). (After Miller and Keen, 1960, and Wickman and Eaton, 1962.)

these conditions has proven to be effective in reducing losses from the western pine beetle (Fig. 8.3).

Individual tree risk assessment methods (Chapter 7) are extremely useful where selection cutting is practiced, permitting the manager to remove those trees that are at greatest risk from insect attack. In areas in which insect problems are particularly severe, selection of trees for cutting may be heavily dependent on their risk characteristics; e.g., in the Eastside Sierra Nevada ponderosa pine forests (Fig. 8.3). Even when formal risk models are not available, the removal of low-vigor trees will usually improve the resistance of stands to insects and, in addition, increase their rates of wood production.

In recent years cutting practices in Europe have evolved to less rigid specifications. Rather than *a priori* decisions to practice evenaged or unevenaged silviculture, cutting decisions are based more on the ecological realities of a given situation and the economic desires of the future. In "free-style" silviculture, as it is sometimes called, each tree is assessed for its functional role in contributing to the desired structure, stability, and social value of the forest. Trees that do not carry any useful function into the future are selected for harvest. Thus, although trees are harvested on individual merit, as in the selection method, different-sized groups or even whole stands may be cut if the trees are growing in nonfunctional groups or stands. In this way, silvicultural prescriptions are made spontaneously in response to current site and stand conditions and the result is a mosiac of unevenaged and evenaged stands intimately adapted to the prevailing ecological and economic conditions, with emphasis on the former.

The key to "free-style" silviculture is identifying the function carriers, those individual trees that will perform a useful future function in stand development and stability. Trees may function in a number of ways, including (1) providing future growing stock (seed producers and understory trees waiting to be released); (2) providing protection for, and regulating the growth rate of, understory trees (overstory trees that reduce light penetration, wind and temperature); and (3) improving the resistance of the stand to insects, diseases, windthrow, and other destructive agents (young, vigorous individuals or resistant species and genotypes). The objective of "free-style" silviculture is to encourage those trees that will function in the development of a future stand with desirable ecological and economic characteristics and that will maintain the stability of that stand. It requires an intimate knowledge of the silvics of the plant species as well as their ecological relationships to each other, to the herbivores that feed on them, and to higher trophic levels (natural enemies of herbivores and of the predators of herbivores).

As we have seen, harvesting has a profound effect on the structure and composition of forest stands, and through this, on future insect populations. In general, cutting should be geared to the silvical characteristics of the tree species being regenerated. Some species, particularly intolerant pioneers, grow better in evenaged stands and may suffer from sunscald, windthrow, and such if they are opened by selection cuttings. The general rule for producing insect resistant stands is to regenerate healthy forests by adapting the harvesting methods to the species and site in question. The structure of the uncut virgin forest can give some clues to the species mixtures and age structures which will develop naturally in a given region. Cutting strategies designed to maintain these natural structures and compositions will usually produce more insect-resistant forests. Cutting can also be used to remove individual trees, or whole stands, before they are attacked by insects. In this way, the food supply for the insects is kept at a low level which prevents their populations from building up. Naturally, accurate risk assessment models are invaluable aids in deciding which trees or stands to cut.

Cutting methods that maintain the diversity of stands will generally decrease their favorability for pest insects. In evenaged stands, diversity can be increased by encouraging a variety of species, whereas pure stands can be diversified by selection cutting to generate an unevenaged structure. The reproduction and survival of pest insects is usually reduced in more diverse ecosystems because the insects have more difficulty in locating susceptible species or age classes when they are mixed among resistant trees. In addition, diverse forest stands usually have a richer fauna of predaceous insects and birds, and this will tend to reduce populations of herbivorous insects. As always, there are exceptions wherein insect problems are aggravated by increasing stand diversity. For example, stands composed of Douglas-fir and spruce are more heavily damaged by the

Cooley gall aphid because the aphid requires both hosts to complete its life cycle, and multistoried unevenaged stands seem to be more susceptible to some defoliators (see Chapter 7, Table 7.2g).

8.1.1.5. Thinning

This method is employed to control stand density so that biomass accumulates on fewer stems. This procedure not only increases the quality and value of timber at harvest but also tends to maintain the stand in a vigorous insect-resistant state (Table 8.1). Thinning also presents the forest manager with an opportunity to remove trees selectively from evenaged stands on the basis of their individual merits. In this way, it allows the forest manager to incorporate some of the desirable features of selection cutting into evenaged management. For example, damaged, diseased, or insect-susceptible trees can be selectively removed during thinning operations, particularly if individual tree risk assessment models are available. Species mixtures can also be altered, perhaps to a more insect-resistant composition, by selectively cutting the more susceptible species.

Like other cutting methods, thinning operations should be planned with the silvical characteristics of the tree species, as well as the entomological and pathological implications, in mind. Some tree species are particularly sensitive to exposure, and opening the stand may result in sunscald or windthrow and, consequently, insect infestations in the weakened trees. Thinning may also increase the incidence of root-rot pathogens that gain entry into the cut stumps and

TABLE 8.1
Mortality and Net Growth of Evenaged Ponderosa Pine Stands in Eastern Oregon 5 Years after Thinning to Different Spacings[a]

Thinning treatment	Stand basal area[b]		Mortality BA[b]		Net growth BA[b]
	1967	1972	Bark beetles	Other	
Unthinned	173.2	152.5	15.8	5.2	−20.7
	(39.76)	(35.01)	(3.63)	(1.19)	(−4.75)
12-foot spacing	116.8	113.5	5.5	0.5	−3.3
	(26.81)	(26.06)	(1.26)	(0.11)	(−0.76)
15-foot spacing	85.8	89.0	0.2	0.3	3.2
	(19.7)	(20.43)	(0.05)	(0.07)	(0.73)
18-foot spacing	61.8	64.8	0.5	0.3	3.0
	(14.19)	(14.88)	(0.11)	(0.07)	(0.69)
21-foot spacing	35.0	37.2	0.0	0.8	2.2
	(8.03)	(8.54)	(0.00)	(0.18)	(0.51)

[a]After Sartwell and Stevens (1975).
[b]Basal area (BA) in square feet per acre and (square meters per hectare).

then move to adjacent trees *via* root grafts or injuries. These root diseased trees may later be attacked and killed by bark beetles. Where root diseases are a problem, it may be necessary to treat cut stumps with fungicides.

8.1.1.6. Timing of Cutting

Losses from insect infestations, particularly in evenaged forests, can be critically influenced by timing of cutting. As a general rule, harvest cuts should be made before individual trees or stands reach ages at which they become susceptible to insect attack. From the entomological standpoint, short *rotations* are therefore preferable because they reduce the time during which trees are available as food for insects and also eliminate the older, less vigorous, age classes. Although the rotation for pulp species may be quite short, timber species may require longer rotations to acquire maximum value. In the latter cases insect populations should be carefully monitored as the stands near maturity.

Besides rotation lengths, the manager also has to consider the time of the year in which harvest cuts or thinnings are made. If logging is carried out at times when insects are inactive, bark beetle infestations of slash can be avoided, damaged trees may have time to recover to full vigor, and infestation of cut logs by wood borers can be eliminated. Because insects are inactive during the cold months, logging in fall and winter is most desirable from the entomological standpoint, provided that the logs are removed and the slash disposed of before springtime. Wood-staining ambrosia beetles are a particular problem when logs are cut in fall and left in the woods until the following spring.

8.1.2. Pruning

Pruning, or the removal of lower crown branches, is a practice designed to improve timber quality by increasing the amount of clear, knot-free, wood. Pruning dead branches has little effect on insect populations but the removal of living branches may increase infestations by pitch moths, resin midges, and some bark beetles attracted by the resin exuding from wounds. Feeding by the larvae of pitch moths and resin midges aggravates the wound and slows the healing process; this may result in timber defects or a fire hazard due to extensive pitch accumulations. Turpentine beetles are also attracted by resin flow from wounded pines. These bark beetles attack the base of the tree and, although they rarely kill their host, they reduce its vigor and make it more susceptible to other tree-killing bark beetles. Some tree species are particularly sensitive to live crown pruning and, of course, heavy pruning of the live crown will affect the vigor of most species. Consideration should always be given to the effect of pruning on the vigor of the tree and, consequently, its susceptibility to insect attack.

8.1.3. Fertilization

Fertilization is used to increase the growth and vigor of forest stands growing on nutrient impoverished sites. Nitrogen, for example, may limit conifer growth on certain sites in western North America, particularly when nitrogen-fixing plants (e.g., alder) have been removed by herbicides to hasten the succession of more valuable conifers (e.g., Douglas-fir). Potassium, phosphorus, calcium, magnesium, and sulfur may also be limiting in certain regions and, in some cases, trace elements such as boron, zinc and manganese may be suboptimal for plant growth. The choice of fertilizer should therefore be based on an analysis of nutrients available in the soil and the requirements of the particular tree species.

Fertilization can have both beneficial and adverse effects on insect pest populations. In many cases fertilization results in a reduction of pest populations through its effects on tree vigor and resistance. Thus, most studies show lower populations of bark beetles and defoliators in fertilized plots. On the other hand, insects limited by nitrogen in their diets (e.g., aphids and scales) may benefit from nitrogen fertilization and, in France, taller fertilized trees tend to be more heavily damaged by processionary caterpillars. The addition of trace elements may also have conflicting effects on insects, on the one hand improving tree vigor, while on the other supplying elements essential for insect reproduction and survival.

8.1.4. Prescribed Burning

Controlled fires are usually employed to reduce fuel loads, reduce understory competing vegetation, and speed up nutrient cycling. These treatments can improve tree growth and insect resistance by increasing the availability of soil nutrients and water. Fires may also directly reduce certain insect populations, such as some *Ips* bark beetles and many defoliators that hibernate in the litter on the forest floor. But burning may destroy the habitat of insectivorous mice and shrews, and removal of the understory vegetation can adversely affect insect predators and parasites that rest or shelter among these plants or that use their nectar as an energy source.

8.1.5. Herbicides

Herbicides are used on commercial forest lands to remove undesirable plants from competition with more valuable species. Although this practice speeds up the succession to the more valuable species, it can also have some undesirable impacts. For instance, herbicides can affect nutrient flows in the soil by removing nitrogen-fixing plants or disrupting soil microorganisms and insects. They may also directly influence pest or natural enemy populations, not to mention their potential effects on human health.

Herbicides often interfere with the natural pattern of ecological succession. The plant species occupying a particular site at a particular time generally reflect the condition of the site, nitrogen-fixing plants such as alder being able to outcompete other species on nitrogen impoverished soils. As nitrogen accumulates, however, other species like Douglas-fir are able to grow and eventually overtop the original pioneer species. Interrupting this natural succession with herbicide treatments may necessitate fertilizer applications at a later date.

The application of herbicides to forest ecosystems can cause complex problems similar to those that result from insecticides (see Chapter 9). The ecological impacts of herbicides, however, are not well understood. It is possible that they can have direct or indirect effects on insect pest populations by changing vegetational diversity, affecting natural enemy populations, or influencing the reproduction and survival of the pests themselves. In addition, herbicides can also pose a hazard to humans and other animals utilizing forest resources. Until the ecological impacts of herbicides are better understood, they should be used with extreme caution.

8.1.6. Prescribed Insect Outbreaks

In addition to herbicides, the silviculturist has another seldom-used tool for suppressing undesirable plant species—the herbivorous insects that feed on them. This approach, commonly referred to as the biological control of weeds, has been used quite frequently to suppress exotic rangeland weeds but never, to my knowledge, native forests trees. There seems to be no theoretical reason, however, why outbreaks of insects feeding on undesirable forest plants cannot be created by management manipulations that improve the environment for these species, such as causing stress to the weed plant and suppressing the natural enemies of the herbivore. One can imagine future silviculturists prescribing stand treatments that improve the environment for the alder flea beetle with the objective of causing severe defoliation to the alder overstory and releasing Douglas-fir in the understory. Encouraging insect outbreaks in this manner not only increases light penetration to the understory but should also increase the rate of nutrient release as insect bodies and feces accumulate on the forest floor. The use of insects as a tool to manipulate forest stands for the benefit of humans is an area that has not received the attention it deserves. Silviculture, after all, should consider all the components of forest ecosystems in achieving its desired objectives.

8.1.7. Silviculture and Natural Enemies

Silvicultural practices can influence the environment of predators, parasites, and pathogens that attack forest insect pests (e.g., Table 8.2). If these natural enemies are important as regulators of pest population densities, environmental

TABLE 8.2

**Effects of Several Cutting Practices on the Composition of Breeding Birds
in Ponderosa Pine–Gambel Oak Forests of Arizona[a]**

	Untreated control	Basal area thinning[b]	Strip shelterwood[c]	Heavy thinning[d]	Clearcut
		Stand structure			
Trees/acre	262	96	74	28	0
Percentage pine	90	91	79	87	0
Percentage oak	8	8	20	13	0
		Nesting guilds[e]			
Cavity and depression nesters	50	59	48	24	1
Foliage nesters	37	67	74	46	12
Ground nesters	23	21	21	14	2
Total numbers	110	147	144	84	15
Number of species	21	22	23	19	6

[a]After Szaro and Balda (1979).

[b]Stands made up of trees less than 10 in. DBH thinned to BA 60, while stands made up of trees greater than 12 in. DBH thinned to BA 70/acre.

[c]Irregular strip shelterwood consisted of clearcut strips (60 feet wide) with most of the oak remaining, interspersed with uncut strips (averaging 120 feet wide) and irregular uncut spacers every 400 feet or so.

[d]Very heavy thinning to BA 22/ft^2 per acre.

[e]Estimated number of nesting pairs per 40 hectares.

changes that improve their habitats but that are not beneficial to the pest will usually result in smaller pest populations and more persistent control. On the other hand, silvicultural practices that reduce the favorability of the environment for natural enemies will lower their effectiveness and increase the likelihood of pest outbreaks. For example, removal of dead snags for fire control purposes reduces the number of nesting sites available to woodpeckers and other hole-nesting insectivorous birds, and controlled burning and grazing can destroy the habitat of predaceous mice and shrews. As avian and small mammal predators are sometimes important in regulating insect populations at low densities (Chapter 5), environmental manipulations of this kind may be instrumental in precipitating pest outbreaks.

Silvicultural practices can also affect populations of insect predators and parasites. Many of these organisms rest on, or find shelter in, understory vegetation, and some utilize the nectar of flowering plants as a source of energy. Practices such as burning and grazing, which remove these plants, may have indirect effects on pest populations. Once again, maintaining the diversity of tree species, ages, snags, and undergrowth will generally produce a more diverse complex of natural enemies and more stable pest populations.

8.1.8. Silvicultural Prescriptions

Silvicultural practices, as we have seen, can have important influences on the dynamics of forest insect populations. For this reason the silviculturist should carefully consider the potential effects of silvicultural activities on insect populations during the preparation of stand prescriptions. Although each specific situation will have its own particular insect problems, some general principles apply under most conditions.

First and foremost is the observation that many insect problems are associated with stressed or unhealthy stands. To minimize this problem, there are several approaches the silviculturist should attempt in an effort to maintain vigorous growing conditions:

1. Selecting tree species and genetic stock for regeneration that are optimally adapted to the site
2. Employing harvest methods that are adapted to the silvics of the tree species being regenerated, often those that mimic the natural pattern of mortality
3. Practicing forest hygiene by removing diseased and unhealthy trees and logging debris, and minimizing damage to standing trees and the site
4. Encouraging diversity in species composition, age structure, and other environmental factors wherever possible
5. Using thinning, fertilization, prescribed fire, and other stand improvement practices to increase vigor and diversity
6. Preventing the accumulation of overmature and senescent trees by setting appropriate rotations.

Second, the silviculturist should consider the biology and ecological requirements of particular insect pests in order to manipulate the environment to their disfavor. For example, certain tip and shoot insects and stem-boring round-headed and flatheaded borers prefer open sunny environments. Where these insects are expected to be a problem, damage can sometimes be avoided by using shelterwood or selection cuts.

Third, the silviculturist should consider the effects of stand treatments on the beneficial arthropod and vertebrate populations. Conditions that improve the environment for birds, small mammals, insect predators and parasites, spiders, and pathogens, will help regulate pest populations at sparse densities. In addition, herbivorous insects that feed on undesirable plants may be used in the control of weed species.

Finally, the job of the silviculturist can be simplified if he or she has access to reliable risk assessment models. With these tools the silviculturist can set priorities for harvest scheduling, choose which trees to mark for selection cutting and thinning, and design appropriate cutting and regeneration methods.

8.2. MANAGEMENT OF NATURAL ENEMIES

As we saw in Chapter 5, different groups of natural enemies play different roles in regulating insect pest populations (Fig. 5.6). Vertebrate predators, and some arthropod parasitoids and predators, seem to be capable of regulating their prey at very low densities. By contrast, pathogenic organisms seem to be more important in suppressing pest outbreaks after they have reached high densities. Between these extremes, arthropod predators and parasitoids may regulate pest populations at low or intermediate densities or help suppress incipient outbreaks. Because each natural enemy group has an important role in the regulation of pest populations at different times and in different places, it is important for the forest manager to try and preserve the variety and diversity of the natural enemy complex in the forest stands. This can best be done by encouraging diversity in the plant community, thereby providing habitats for different kinds of insectivorous organisms. *Preservation* of natural enemy populations should be practiced wherever possible, and particularly during treatments with chemical pesticides. For example, insecticidal sprays sometimes reduce arthropod predator and parasitoid populations more severely than the target pest, and some of them may affect the reproduction and survival of birds and mammals as well (e.g., the "thin eggshell" syndrome caused by the insecticide DDT). Predator and parasite populations can sometimes be preserved during chemical treatment by timing applications to periods when they are not exposed to the spray or avoiding areas where they concentrate (Table 8.3).

Foresters interested in preventing insect outbreaks should pay particular attention to predators capable of regulating pest populations at low densities. Vertebrate predators and insect parasitoids that have sigmoid functional responses or very rapid numerical responses seem to possess this capability. What then can the forester do to improve the effectiveness of these natural enemies? In Chapter 5, we saw that predators with sigmoid functional responses are only able to regulate their prey at low densities if their own populations are sufficiently large (Fig. 5.3d). We also know that, everything else being equal, the rapidity of the predator numerical response is dependent on the initial number of predators present. The regulatory efficiency of these natural enemies can be improved, therefore, by increasing the density of their populations.

Augmentation of parasite and predator populations is a useful preventive technique, but its success depends on an understanding of the forces that limit natural enemy populations. Obviously, if a predator population is limited by the density of the pest species, the manager can do little besides rearing the predators artificially and releasing them into the field in large numbers. Augmentation by natural enemy release, however, is unlikely to result in long-term regulation because food-limited predator populations will automatically adjust to their original density as determined by their food supplies. Strategic natural enemy releases

TABLE 8.3
Examples of Natural Enemy Preservation, Augmentation, and Introduction for the Control of Forest Insect Pests

Pest species	Technique and countries	Source
Nun moth Oak processionary Gypsy moth	*Preservation* of insect parasitoids during chemical control operations by spraying prior to adult parasitoid flight (Europe)	Franz (1961)
Western pine beetle	*Preservation* of clerid predators by leaving tree stumps and surrounding litter, where predators accumulate, untreated during chemical control operations (USA)	Berryman (1967)
Great spruce beetle	*Augmentation* of an insect predator by field releases	Grégoire *et al.* (1985)
Pine processionary	*Augmentation* of red forest ant colonies by relocation from other sites (Europe)	Franz (1961)
Pine sawfly Pine caterpillar	*Augmentation* of insect parasitoids by repeated field releases (Europe)	Franz (1961)
Defoliators	*Augmentation* of cavity-nesting birds 5- to 10-fold by supplying nesting boxes (Europe and Russia)	Franz (1961)
Larch sawfly	*Augmentation* of deer mice by providing nesting boxes, and red-backed voles and masked shrews by piling brush for nesting sites (USA)	Buckner (1966)
Gypsy moth	*Augmentation* of white-footed mice by providing nesting boxes and alternative food, and by removing protected hiding places for gypsy moth eggs, larvae, and pupae (USA), and augmenting natural enemies by supplemental feeding with release of gypsy moth egg masses (Europe)	Smith and Lautenschlager (1981); Campbell *et al.* (1975, 1976); Maksimovic *et al.* (1970) (Fig. 8.4b)
Larch sawfly	*Introduction* of parasitoids from Europe and the masked shrew from the mainland into Newfoundland (Canada)	Turnbull and Chant (1961); Buckner (1966)
Balsam woolly adelgid	*Introduction* of beetle predators from Europe into Canada and the USA	Turnbull and Chant (1961)
Larch casebearer	*Introduction* of parasitoids from Europe into Canada and the USA	Turnbull and Chant (1961)
Spruce sawfly	*Introduction* of parasitoids and a virus (accidental) from Europe into Canada	Turnbull and Chant (1961)
Winter moth	*Introduction* of parasitoids from Europe into Canada (see Fig. 5.10)	Embree (1966)
European woodwasp	*Introduction* of parasitoids and a parasitic nematode from Europe into Australia	Taylor (1982)

may be effective in reducing incipient outbreaks, however, by reducing the time delay in the predator's numerical response (Fig. 8.4a). It may also be possible to limit the intensity or extent of pest outbreaks by artificially increasing the food supply of natural enemies at times when prey are scarce; i.e., by augmenting the prey or pest population (Fig. 8.4b). This experiment illustrates how peak pest densities can be reduced by supplementing the food supply of natural enemies.

Many natural enemies are limited by factors other than the density of a particular pest species. Cavity-nesting birds, for example, may be limited by the abundance of dead snags in which to build their nests, and mice and voles may be limited by nesting places or by food shortages at times when their insect prey are unavailable; i.e., insects are only available to most rodents when they are on the forest floor. Thus, populations of insectivorous vertebrates can sometimes be augmented by habitat management. For example, leaving dead snags or providing nesting boxes for birds and mice, leaving piles of debris (slash) for shrew and mouse nesting sites, or providing supplementary diets during times of food shortages (Table 8.3). Arthropod predators and parasitoids can also be augmented by habitat management, particularly by improving undergrowth diversity, which provides them with places to rest, hide, and a secondary source of food, such as nectar and other prey species.

Relocation is another way in which arthropods and vertebrates can be augmented. In Europe, for example, ant colonies are sometimes moved into regions where pest outbreaks are expected (Table 8.3). Finally, the effectiveness of natural enemies can sometimes be augmented by making the prey more vulnera-

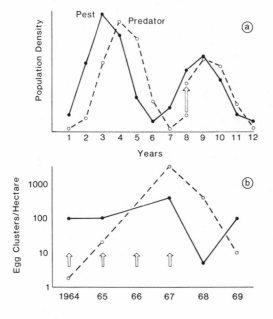

FIGURE 8.4. (a) Reduction of an incipient pest outbreak by strategic release of natural enemies, which reduces the time delay in the numerical response (hypothetical situation). Arrow indicates time of release. (b) Reduction of gypsy moth peak densities by adding egg masses to the population in order to supplement the food supply of natural enemies. Broken line = untreated population; solid line = treated population; arrows = release of gypsy moth egg masses. (After Maksimovic *et al.*, 1970.)

ble to attack. For example, gypsy moth larvae in sparse populations usually rest during the day, and then pupate under bark flaps, signs nailed to trees, and other sites protected from vertebrate predators. The removal of these hiding places, many of which are the result of human activities, may significantly increase the efficiency of vertebrate predators in maintaining sparse and stable gypsy moth populations (Table 8.3).

It is important to realize that the regulation of pest populations by natural enemies with sigmoid functional responses, but with slow or nonexistent numerical responses (e.g., vertebrate predators), may be rather tenuous. This is because unstable outbreak thresholds are created when the functional responses saturate (see Fig. 5.3b). Regulation of pest populations by these agents can be disrupted by immigrations of the prey species or by alterations in the favorability of the environment. The resulting pest eruption can then spread by immigration over large forested areas (see Fig. 4.10). From this perspective, arthropod predators and parasitoids, with their potential for rapid numerical responses, may be much more effective as permanent regulating agents.

A special kind of problem arises when an insect pest is accidentally introduced into a new environment, usually from a different continent. Not only are these insects exposed to tree species that have not adapted genetically to their feeding, but they often arrive without their normal complement of natural enemies. Because of this they usually increase to extremely high densities and quickly spread throughout susceptible forests. What is created by a pest introduction is, in fact, an eruptive pest outbreak that may settle into a permanent outbreak condition. One solution to this problem that did not escape early forest entomologists is to introduce the native coadapted parasites and predators of the pest with the hope that they will regulate the pest population at low, nondamaging, levels. Natural enemy *introduction,* known popularly as biological control, has been attempted with practically all exotic forest pests. On the whole, about half these attempts have been successful or partially successful in controlling the introduced pest.

The introduction of organisms of any kind into a new country is governed by strict quarantine laws and, therefore, a specific protocol is normally followed when natural enemy introductions are planned. This protocol usually involves nine steps and can be associated with the acronym IRIS-SIRME:

1. *Identifying* the pest species and its native home
2. *Reviewing* all information, both published and verbal, concerning the pest and its natural enemies
3. *Inventorying* natural enemy species from the literature, observation in the field, and laboratory rearings of parasitized insects
4. *Studying* the effectiveness of various natural enemies at regulating pest populations in the native home, their host specificity and alternative hosts, and their life histories, fecundities, host-finding behaviors, dis-

persal abilities, and whether they are ecologically adapted to the new environment

5. *Selecting* the most promising natural enemies for importation based on their potential for regulating the pest in its new home, and *screening* these species to make sure they do not attack beneficial insects or plants

6. *Importing* selected natural enemies and rearing them in large numbers on their host under strict quarantine control, at which time attempts are made to eliminate hyperparasites and diseases of the natural enemies

7. *Releasing* natural enemies into the new environment. (A decision must be made whether to release one or more enemies simultaneously or in sequence. There is some controversy over the best release strategy but a common compromise is to release the two or three most promising species one at a time. Releases should be made in areas of high host density and favorable climatic and habitat conditions.)

8. *Monitoring* the buildup and spread of the natural enemy populations by periodic sampling

9. *Evaluating* the success of the introduction in terms of pest population regulation at low, subdamaging levels over a relatively long period of time

It may be significant that most of the successful attempts at biological control of exotic pests have involved insect parasitoids. This is partly because these insects have effective host-finding behaviors, good dispersal abilities, and rapid reproductive numerical responses and partly because of the preferences of biological control specialists. However, predaceous beetles and mammals, virus pathogens, and nematodes have all contributed to the regulation of introduced pests (see Table 8.3).

Obviously, one of the most effective ways to prevent pest introductions is by quarantining or embargoing infested materials. Most countries have governmental quarantine and inspection services whose purpose is to examine all imported plant material for the presence of pest organisms. In this way, large numbers of pests are intercepted before they are able to enter a new habitat. In addition, many countries have laws forbidding the importation of certain plants and plant products, particularly when major agricultural or forest crops are threatened by specific pests. In spite of all these regulations, it is impossible to prevent all potential pest introductions. Too many humans and their commercial products move continuously over too many traffic lanes. Quarantines and embargos may slow but cannot stop the movement of pest organisms around the world.

REFERENCES AND SELECTED READINGS

Balch, R. E., 1957, Control of forest insects, *Annu. Rev. Entomol.* **3**:449–468. (reviews biological and silvicultural control)

Baumgartner, D. M., and Mitchell, R., (eds.), 1984, Silvicultural management strategies for pests of the interior Douglas-fir and grand fir forest types, Cooperative Extension, Washington State University, Pullman.

Belanger, R. P., and Malac, B. F., 1980, Silviculture can reduce losses from the southern pine beetle, U.S. Forest Service, Agriculture Handbook No. 576, 17 pp.

Berryman, A. A., 1967, Preservation and augmentation of insect predators of the western pine beetle, *J. For.* **55**:260–262.

Berryman, A. A., 1982, Biological control, thresholds, and pest outbreaks, *Environ. Entomol.* **11**:544–549.

Buckner, C. H., 1966, The role of vertebrate predators in the biological control of forest insects, *Annu. Rev. Entomol.* **11**:449–470. (excellent review of the potential of vertebrates for controlling forest insects)

Campbell, R. W., Hubbard, D. L., and Sloan, R. J., 1975, Location of gypsy moth pupae and subsequent pupal survival in sparse, stable populations, *Environ. Entomol.* **4**:597–600.

Campbell, R. W., Miller, M. G., Duda, E. J., Biazak, C. E., and Sloan, R. J., 1976, Man's activities and subsequent gypsy moth egg-mass density along the forest edge, *Environ. Entomol.* **5**:273–276. (man-made junk may increase the likelihood of outbreaks)

Dowden, P. B., 1962, Parasites and predators of forest insects liberated in the United States through 1960, U.S. Forest Service Agriculture Handbook No. 226, 70 pp.

Embree, D. G., 1966, The role of introduced parasites in the control of the winter moth in Nova Scotia, *Can. Entomol.* **98**:1159–1167.

Franz, J. M., 1961, Biological control of pest insects in Europe, *Annu. Rev. Entomol.* **6**:183–200. (general review of biological control)

Garton, E. O., and Langelier, L. A., 1985, Effects of stand characteristics on avian predators of western spruce budworm, in: The Role of the Host in the Population Dynamics of Forest Insects, Proceedings of the IUFRO Conference, Banff, Canada (L. Safranyik, ed.), pp. 56–72, Pacific Forest Research Centre, Victoria, B.C.

Grègoire, J.-C., Merlin, J., Pasteels, J. M., Jaffuel, R., Vouland, G., and Schvester, D., 1985, Biocontrol of *Dendroctonus micans* by *Rhizophagus grandis* Gyll. (Col., Rhizophagidae) in the Massif Central (France), *Z. Angew. Entomol.* **99**:182–190.

Keen, F. P., 1946, Entomology in western pine silviculture, *Pan-Pacific Entomol.* **22**:1–8. (classic review of silvicultural approaches to insect control)

Keen, F. P., and Craighead, F. C., 1927, The relation of insects to slash disposal, U.S. Department of Agriculture Circular No. 411, 12 pp.

Lejeune, R. R., McMullen, L. M., and Atkins, M. D., 1961, The influence of logging on Douglas-fir beetle populations, *For. Chron.* **37**:308–314.

Maksimovic, M., Bjegovic, P. L., and Vasiljevic, L., 1970, Maintaining the density of the gypsy moth enemies as a method of biological control, *Zastita Bilja* **107**:3–15.

McClelland, B. R., Frissell, S. S., Fischer, W. C., and Halvorson, C. H., 1979, Habitat management for hole-nesting birds in forests of western larch and Douglas-fir. *J. For.* **77**:480–483. (bird density highest in old growth stands)

Miller, J. M., and Keen, F. P., 1960, Biology and control of the western pine beetle, U.S. Forest Service Miscellaneous Publication No. 800, 381 pp.

Mitchell, R. G., Waring, R. H., and Pitman, G. B., 1983, Thinning lodgepole pine increases tree vigor and resistance to mountain pine beetle, *For. Sci.* **29**:204–211.

Otto, Von H.-J., 1985, Sylviculture according to site conditions as a method of forest protection, *Z. Angew. Entomol.* **99**:190–198.

Prebble, M. L., 1951, Forest entomology in relation to silviculture in Canada, *For. Chron.* **27**:1–32. (also includes papers by R. E. Balch and G. W. Barter, M. L. Prebble and R. F. Morris, R. R. Lejeune, G. R. Hopping, H. A. Richmond, and J. M. Kinghorn on specific insects)

Pschorn-Walcher, H., 1977, Biological control of forest insects, *Annu. Rev. Entomol.* **22**:1–22. (good review of natural enemies in pest control)

Safranyik, L., Shrimpton, D. M., and Whitney, H. S., 1974, Management of lodgepole pine to reduce losses from the mountain pine beetle, Environment Canada, Forest Technical Report No. 1, 24 pp.

Sartwell, C., and Stevens, R. E., 1975, Mountain pine beetle in ponderosa pine—Prospects for silvicultural control in second-growth stands, *J. For.* **73:**136–140.

Schmid, J. M., Thomas, L., and Rogers, T. J., 1981, Prescribed burning to increase mortality of pandora moth pupae, U.S. Forest Service Research Note RM-405, 3 pp.

Schmiege, D. C., 1963, The feasibility of using a neoaplectanid nematode for control of some forest insect pests, *J. Econ. Entomol.* **56:**427–431.

Smith, H. R., and Lautenschlager, R. A., 1981, Gypsy moth predators, in: The gypsy moth: Research towards integrated pest management. (C. C. Doane and M. L. McManus, eds.), U.S. Forest Service Technical Bulletin 1584, pp. 96–125.

Stark, R. W., 1965, Recent trends in forest entomology, *Annu. Rev. Entomol.* **10:**303–324. (reviews forest fertilization effects on insect populations)

Syme, P. D., 1975, The effects of flowers on the longevity and fecundity of two native parasites of the European pine shoot moth in Ontario, *Environ. Entomol.* **4:**337–346.

Szaro, R. C., and Balda, R. B., 1979, Effects of harvesting ponderosa pine on nongame bird populations, U.S. Forest Service Research Paper *RM-212,* 8 pp.

Taylor, K. L., 1982, The sirex woodwasp: Ecology and control of an introduced forest insect, in: *The Ecology of Pests: Some Australian Case Histories* (R. L. Kitching and R. E. Jones, eds.), CSIRO, Australia, pp. 231–248.

Turnbull, A. L., and Chant, D. A., 1961, The practice and theory of biological control of insects in Canada. *Can. J. Zool.* **3:**697–753. (classic review of forest pests control with natural enemies)

USDA, 1977, Biological Agents for Pest Control—Status and Prospects, Superintendent of Documents, U.S. Government Printing Office, Washington, D.C., 138 pp.

van den Bosch, R., Messenger, P. S., and Gutierrez, A. P., 1982, *An Introduction to Biological Control,* Plenum Press, New York. (good general introduction to pest control with natural enemies)

Wickman, B. E., and Eaton, C. G., 1962, The effects of sanitation-salvage cutting on insect-caused mortality at Blacks Mountain Experimental Forest 1938–1959, U.S. Forest Service Technical Paper No. 66, 39 pp.

CONTROL OF INSECT OUTBREAKS

In the previous chapter we discussed methods for preventing insect populations from reaching outbreak levels. If these practices fail, or if they are not implemented, the forest manager will probably have to face pest outbreaks from time to time. Once insect populations have attained levels that cause serious damage to forest resources, the manager has three options: (1) do nothing and allow the outbreak to run its course; (2) attempt to limit the spread of the outbreak; and (3) attempt to reduce the insect population to nondamaging levels. In this chapter we will discuss the basic principles of outbreak containment and suppression and then explore the tactics that can be employed to achieve these ends.

9.1. PRINCIPLES OF INSECT PEST CONTROL

There are two active strategies for controlling ongoing pest outbreaks. The first strategy, called containment, attempts to limit the extent of the damage by erecting barriers to population expansion. The strategy of suppression, on the other hand, attempts to reduce the intensity of damage by reducing the pest population to low densities. The success of these strategies depends to a great extent on the population dynamics exhibited by the particular pest. In this section, we will examine the strategies of containment and suppression in the light of general population theory (Chapter 4), to see when and where they are likely to succeed or fail. We will also discuss some of the problems likely to be encountered when using these control strategies.

9.1.1. Containment of Outbreaks

The objective of containment is to prevent or slow down the spread of an insect outbreak into uninfested or lightly infested areas. For this reason, containment is not normally feasible against gradient outbreaks which are restricted to certain susceptible stands and sites and rarely spread into nonsusceptible regions. On the other hand, containment may be a viable strategy against eruptive pests and introduced exotic insects that tend to spread from specific epicenters.

The success of a containment strategy depends on three conditions: (1) that the outbreak region can be clearly identified and delineated; (2) that an effective barrier can be erected around the outbreak region; and (3) that any penetration of the barrier can be rapidly eliminated or contained. Obviously the probability of successful containment is increased if the pest has poor dispersal capabilities or if a highly effective barrier can be created.

Containment strategies are most commonly employed against exotic pests because the epicenter is usually small and well defined, often in the vicinity of ports of entry. Legislative barriers in the form of quarantines and embargoes are often erected to prevent or retard the movement of infested material out of the epicenter. If insects are found outside the quarantined region, attempts may be made to eradicate them. In some cases, pest eradication may also be attempted within the quarantined region.

The containment of native pests is a much more difficult problem. In many cases, the outbreak epicenters are not clearly definable, and effective barriers to dispersal are not available. In addition, many eruptive insect species have well-developed flight abilities or are transported for long distances by air currents and weather fronts. However, containment may be a viable strategy with some aggressive bark beetles such as the southern pine beetle, mountain pine beetle, western pine beetle, and European spruce bark beetle. Outbreak epicenters of these species are easily delineated by the red or brown foliage of infested trees, and effective barriers can sometimes be erected using the aggregation pheromones of the insects. In this case, living trees within the epicenter, or around the perimeter of the epicenter, are baited with aggregation pheromones. Beetles emerging from dead trees then attack the baited trees as well as neighboring individuals, so that the infestation remains within the general vicinity of the epicenter. Containment of aggressive bark beetle populations, however, must be considered a temporary policy aimed at maintaining the insect population within areas awaiting harvest. Similar techniques using aggregation pheromones can also be employed to shift beetle populations into regions scheduled for immediate harvest.

9.1.2. Suppression of Outbreaks

The aim of suppression is to reduce damaging insect outbreaks to tolerable levels. One of the problems is to decide on what population density, or level of damage, is intolerable. The solution to this problem is no trivial task and is addressed in more detail in the next chapter. For the present, however, let us consider the intolerable population to be that which exceeds the *economic damage level* (EDL), the point at which damage to forest resources exceeds the costs of controlling the outbreak. Apart from knowing when to apply suppression, the manager should also understand how suppressive treatments are likely to affect different kinds of insect outbreaks.

9.1.2.1. Suppression of Gradient Outbreaks

Gradient outbreaks are often associated with particular site conditions, so that populations are consistently high on susceptible sites but not on others (Fig. 9.1a). When these outbreaks are suppressed, they automatically return toward their equilibrium levels and require repeated treatments to maintain them below the EDL (Fig. 9.1a). Repeated application of suppressive treatments can be costly, however, particularly when computed over the rotation period. For this reason, they can rarely be justified on economic grounds in most forestry situations. Rather, the site should be converted to an alternative use or planted with a less susceptible tree species.

On the other hand, suppression may be more feasible against insects that exhibit temporary pulse gradients. These outbreaks are associated with environ-

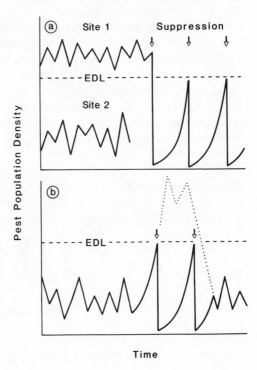

FIGURE 9.1. (a) Suppression of sustained gradient outbreaks associated with spatial variations in the environment (site 1, favorable environment for the pest; site 2, unfavorable environment) results in the "rebound" of the pest population back toward its equilibrium density. (b) Pulse-gradient outbreaks associated with time-varying environmental factors can be suppressed below EDL with a few well-timed treatments.

mental variables that vary with time—droughts, gales, outbreaks of other insects or pathogens—and they subside when the environment returns to normal. Temporary outbreaks of this nature can be suppressed with one or a few judiciously timed treatments (Fig. 9.1b).

Cyclical gradient outbreaks are caused by delayed negative feedback between the pest and its host plant or natural enemy populations. If the causal mechanisms are known or suspected, suppression strategies can be designed that have a good chance of controlling the pest population below its EDL. For instance, larch budmoth population cycles in the European Alps are thought to be caused by delayed defensive reactions in the foliage of their host plants; i.e., the foliage of the defoliated larch trees is shorter, tougher, covered with resin, and contains lower concentrations of nutrients for 2 or 3 years after heavy defoliation. If this hypothesis is true, one would not expect suppression treatments to have any dramatic effects on the amplitude or period of the cycle. This conclusion is supported by some experimental suppression experiments against budmoth populations in the French Alps (Fig. 9.2). In the first experiment, the budmoth population was suppressed with a bacterial pesticide at the low point of its cycle. This treatment delayed the next population peak by 1 year but did not prevent the population from exceeding the economic damage level (Fig. 9.2a). The second treatment was applied at the peak of the cycle and, although it had no effect on the natural cyclical trajectory, it did reduce the population below the economic damage level during the critical peak year (Fig. 9.2b). As a result, the forest suffered very little damage from this outbreak.

It is apparent from Fig. 9.2 that suppression activities have little influence on the course of larch budmoth cycles, presumably because they do not affect the host–insect interaction. If cyclic outbreaks are dependent on plant–insect interactions, the manager can only alter their course by manipulating the plant population in some way (Chapter 8). Suppression, however, can sometimes be

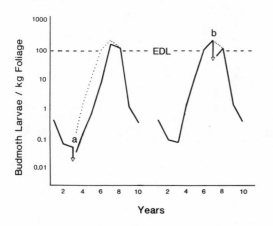

FIGURE 9.2. Experimental suppression of larch budmoth populations with applications of a bacterial pesticide (*Bacillus thuringiensis*). (- - -) Trajectory of untreated populations. (a) Treatment applied at the low point of the cycle reduced the populations by 60% and delayed the cycle peak by 1 year. (b) Treatment applied at the peak of the cycle reduced the population by 80% to below the economic damage level. (Modified from Martouret and Auer, 1977 and Auer *et al.*, 1981.)

used to prevent pest populations from exceeding economic damage levels (Fig. 9.2b).

In certain pest species, cyclical gradient outbreaks may be caused by time delays in the numerical responses of natural enemy populations (Chapter 8). If suppression tactics are used against these pests, and if the treatments also reduce natural enemy populations to the same extent, they may be followed by even worse outbreaks (Fig. 9.3a). If, on the other hand, treatments only affect the pest species and not the natural enemies, then careful application can reduce the amplitude of the next outbreak cycle (Fig. 9.3b). Suppression methods that only affect the pest species are called *selective* or *narrow-spectrum* treatments. It is important to realize that the suppression of pest population cycles with selective treatments depends on obtaining optimal rather than maximal pest mortality. In other words, the pest population should only be reduced to the extent that natural enemies are able to catch up and control the population (Fig. 9.3b). Aiming for maximal kill, which is a more frequent pest-control objective, will often have the same effect as using nonselective treatments because natural enemies will either starve or move out of the area when their prey becomes very scarce.

9.1.2.2. Suppression of Eruptive Outbreaks

Eruptive outbreaks occur when the survival and/or reproduction of pest insects is positively related to the density of their populations over certain ranges

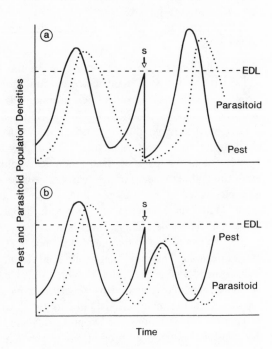

FIGURE 9.3. Hypothetical effect of suppressing a pest population cycle resulting from interaction with a specific insect parasitoid. (a) Nonselective suppression tactic that reduces both parasitoid and pest populations by the same proportion. (b) Selective suppression tactic that only affects the pest population. Note that the pest population density should not be reduced too much as this may cause a decrease in the parasitoid population due to starvation or emigration.

of population density (Chapter 4). Positive feedback can result when cooperative behavior enables the insect population to overwhelm resistant hosts or to escape predators or when the responses of natural enemies to the prey population saturate at high prey density (Chapter 5). Examples can be found among the bark beetles, some of which are able to overwhelm healthy hosts by pheromone-mediated mass attack; among the sawflies, some of which form defensive aggregations that reduce natural enemy attacks; and certain defoliators, some of which are regulated by predators or parasitoids that become ineffective at high prey densities.

The major difference between pest eruptions and the other outbreak classes is that they invariably spread from very susceptible stands (the epicenters) into adjacent less susceptible areas. For this reason, containment may be a viable strategy with these pests (Section 9.1.1).

In certain cases, it may also be possible to terminate eruptive outbreaks by carefully planned suppression treatments. For example, some population eruptions are triggered by temporary environmental disturbances that enable the population to rise above the outbreak threshold (Chapter 4). Once a population has exceeded this threshold, however, it may continue to expand even though the environment returns to normal; e.g., bark beetle and spruce budworm outbreaks may be triggered by dry, warm weather, but then continue unabated for many years even though the weather returns to normal. If these populations can be suppressed below their outbreak thresholds, then mass destruction of the forest can be avoided (Fig. 9.4a).

If this strategy is to work, however, it is essential that the following conditions be met:

1. The forest environment must not be too favorable for the reproduction and survival of the pest. In other words, single suppressive treatments will fail to terminate permanent or cyclical pest eruptions because the environment remains in a very favorable condition (as in the right-hand column of Chapter 4, Fig. 4.12). Under these conditions, eruptive outbreaks behave in a similar manner to gradient outbreaks, with the exception that they tend to expand through space.

2. The pest population must be reduced below the outbreak threshold over the entire outbreak area. For this reason the strategy is more likely to succeed if it is implemented as early as possible, while the affected area is still relatively small. With eruptive outbreaks, therefore, suppression should be applied well before the economic damage level is reached.

3. The suppression treatment does not have an adverse effect on the outbreak threshold. This condition would hold for most cases in which the threshold is determined by plant resistance factors (e.g., bark beetles). However, when the outbreak threshold is caused by predator or parasitoid attack responses, some suppressive tactics may seriously disrupt the threshold. For instance, if predator or parasitoid populations are suppressed by the treatment, then the outbreak threshold will be reduced proportionally (see Chapter 5, Fig. 5.3d).

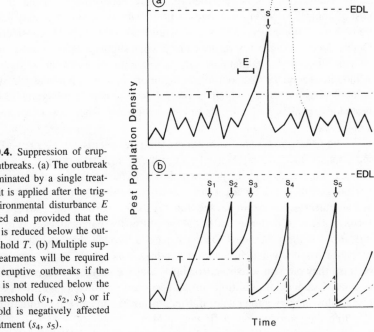

FIGURE 9.4. Suppression of eruptive pest outbreaks. (a) The outbreak can be terminated by a single treatment s, if it is applied after the triggering environmental disturbance E has subsided and provided that the population is reduced below the outbreak threshold T. (b) Multiple suppression treatments will be required to control eruptive outbreaks if the population is not reduced below the outbreak threshold (s_1, s_2, s_3) or if the threshold is negatively affected by the treatment (s_4, s_5).

If any of these conditions are not met, the termination of eruptive outbreaks with single suppression treatments becomes impossible, and suppression to subeconomic damage levels can only be achieved by several, or in some cases many, treatments. For example, multiple treatments are required if they fail to reduce the pest population below the outbreak threshold (Fig. 9.4b: s_1, s_2, s_3) or if the outbreak threshold is seriously impacted by the treatment (Fig. 9.4b: s_4, s_5).

9.1.3. Effects of Treatments on Nontarget Organisms

As mentioned earlier, pest control tactics can be classified according to the degree of selectivity they show towards the pest species. For example, highly selective (narrow-spectrum) tactics such as species-specific pheromones or viruses have a strong suppressive effect on the target insect but usually have little or no direct harmful effects on other components of the forest ecosystem. On the other hand, certain ''broad-spectrum'' chemical pesticides may have a direct impact on a large number of beneficial parasitoids and predators and in some cases even the host plant (i.e., they may be phytotoxic).

The use of broad-spectrum control tactics can precipitate a series of undesirable side effects. Nonselective suppressive treatments may lead to pest population resurgences that may be more damaging than the original outbreak (Fig. 9.3a). These resurgences may then require repeated treatments that increase the cost of control. In addition, broad-spectrum suppression tactics may also induce outbreaks of nontarget species that would not otherwise be considered pests. Outbreaks of *secondary pests* are caused when suppressive treatments disrupt the parasitoids or predators that normally maintain their populations at low densities. Although secondary pest outbreaks are more commonly associated with agricultural ecosystems, in which heavy reliance is placed on synthetic chemical pesticides, similar problems also occur in forestry when pesticides are used in similar ways.

The use of broad-spectrum suppression treatments can also have far-reaching impacts if the materials used do not break down rapidly into harmless substances. Chemicals that break down slowly in the environment are termed *persistent,* and their effects are often transmitted through food webs, waterways, and air currents. For example, the thin-eggshell syndrome that causes the eggs of raptorial birds to collapse under the weight of the parent is caused by *biomagnification* of the pesticide DDT as it passes through the food chain. Biomagnification occurs when insecticide residues concentrate in the top carnivores because they eat many fish and/or birds that themselves have eaten many contaminated insects. In addition, broad-spectrum chemicals can affect soil organisms, slowing down the rate of nutrient turnover and can kill pollinating insects, affecting the reproduction of flowering plants.

Pest-control chemicals can also be transported to far distant ecosystems *via* water and air movements. For example, tuna in the Pacific Ocean have been seriously contaminated by DDT sprayed on agricultural and forest crops hundreds of miles distant. This problem is particularly severe with persistent toxins that have a long life span—the so-called *residual* effect. Thus, whenever broad-spectrum treatments are contemplated, attempts should be made to use materials with short residual effects, e.g., chemicals that break down fairly rapidly in the environment.

Although the vast majority of broad-spectrum chemicals are used in agriculture, forest systems are often much more sensitive to these materials. In comparison with forests, agricultural ecosystems are relatively sterile and crop yields are largely governed by human inputs such as fertilizers, herbicides, and insecticides. By contrast, forest yields rely to a greater extent on natural nutrient cycles and natural regulation of pest populations by predators, parasites, diseases, and host resistance. In addition fish, game, and esthetic values may be considered yields from forest ecosystems. For this reason, forests are much more easily disturbed by the careless use of broad-spectrum persistent pesticides.

9.1.4. The Resistance Problem

Whenever pests are repeatedly subjected to suppressive tactics, it is always possible that they will evolve resistance to that particular control method (Chapter 1). Perhaps resistance is not the best term for this natural process of genetic adaptation to an environment that contains a new hazard. Under these conditions, the most vulnerable genotypes are killed by the suppressive treatment, while the more resistant ones give rise to the next generation. Over a number of successive generations and repeated treatments, a resistant population evolves that can only be suppressed by more concentrated chemical doses or that must be controlled by different tactics. Although resistance is a common phenomenon in agricultural and medical pests, it occurs less frequently in forestry, where continuous chemical treatments are uncommon. Where forests are repeatedly treated with the same material, however, one should expect resistance to develop eventually in the pest population. This has apparently happened in eastern North American forests, which have been sprayed year after year to suppress spruce budworm populations.

9.1.5. Summary of Pest Control Principles

Strategies for containing or suppressing forest insect outbreaks should be applied with caution and foresight and with particular attention to the economic and social costs of the operation and the population dynamics of the particular pest species. In general,

1. Containment strategies are sometimes feasible with exotic insects or with native eruptive pest outbreaks. Successful containment requires that epicenters be clearly delineated, that effective barriers to insect dispersal be erected, and that infestations resulting from penetration of the barrier be eliminated.

2. Suppression of sustained gradient outbreaks requires continuous treatment and is only feasible in high-value crops. Pulse and cyclical gradients, however, can often be suppressed below the economic damage level by one or a few strategically timed treatments.

3. Eruptive outbreaks can sometimes be terminated by single strategically timed suppression treatments, provided the disturbing forces have subsided, the total outbreak population is effectively suppressed below the outbreak threshold, the treatment does not adversely affect the outbreak threshold, and the eruption is not permanent or cyclical.

4. Whenever possible, narrow-spectrum nonpersistent treatments should be used so as to protect beneficial organisms, avoid secondary pest outbreaks, and reduce ecosystem damage.

5. Continuous application of the same suppressive tactic should be avoided or resistance to the treatment will probably evolve in the pest population.

9.2. CHEMICAL INSECTICIDES

Chemical poisons or insecticides, for all their problems, remain the most commonly used weapons for controlling pest outbreaks. For example, 5 billion kg of insecticides was used in the United States alone from 1945 to 1970. In the early days, inorganic poisons such as lead arsenate, calcium arsenate and, on occasion, elemental sulfur were the main chemicals employed. During the 1940s, however, the first synthetic insecticides began to appear on the market, and the success of these materials against the insect vectors of malaria, typhus, and other human diseases, particularly during World War II, gave rise to the huge pesticide industry of today.

Individual insecticides are referred to by a number of different names, sometimes leading to considerable confusion. First, there is the *chemical name*, which is analogous to the scientific name of an organism. A typical chemical name would be *1-naphthyl methylcarbamate* for an insecticide belonging to the carbamate group. Then each material is given an approved *common name* such as *carbaryl* for the aforementioned carbamate. Finally, the insecticide may have one or more registered trade names, depending on how many different companies market the substance. In our example, the trade name is Sevin. Common names will be used most of the time in this chapter.

Insecticides often have to be *registered* by government agencies before they can be used operationally against specific targets. In the United States, for example, the Federal Insecticide, Fungicide, and Rodenticide Act of 1947 and its amendments specifies the procedure for insecticide registration. For registration purposes, the manufacturer is required to provide proof of the insecticide's efficiency in controlling the target organism and that the material does not pose unacceptable hazards to human health or the environment. Surprisingly few insecticides are registered for use against forest insects in the United States.

Chemical insecticides are characterized by chemical properties, origin, mode of action, toxicity to insects and vertebrates, and specificity and persistence. Some of the more common groups of insecticides are discussed below.

9.2.1. The Organochlorines

As the name suggests, organochlorines are characterized by chlorine atoms attached to an organic molecule, usually a benzene ring structure. The first organochlorine insecticide, DDT, was synthesized in 1874, but its insecticidal properties were not discovered until 1939. DDT was used extensively in World War II to control body lice that transmit the dreaded typhus pathogen. After the war, the "miracle" insecticide was used to control the mosquito vectors of malaria and, in the more developed countries, against agricultural and forest insects. Up until the time it was withdrawn from the U.S. market in 1973, most major outbreaks of forest insects in North America were sprayed with DDT. In 1968, for example, 20,000 kg of DDT was deposited on forested lands in the United States to control defoliating insects. Organochlorine insecticides were also used in fairly large quantities to control bark beetles and termites.

The organochlorine insecticides are lipophilic compounds that are absorbed through the cuticle of the insect. For this reason they, and most other organic insecticides, are called *contact poisons* (most inorganic compounds, in contrast, are called stomach poisons because they have to be ingested). Organochlorine poisoning in both insects and mammals causes nerve fibers to produce repetitive discharges in response to a single stimulus. This results in continuous tremors, convulsions, muscle fatigue, and death. In effect, the animal twitches to death.

The organochlorines are generally much more toxic to insects than to other animals. For example, the dosage of DDT which kills 50% of a group of nonresistant houseflies, called the LD_{50}, is 8 mg/kg of body weight, while the LD_{50} for the rat is 118 mg/kg (see Table 9.1).

The widespread and massive use of DDT, and to a lesser extent other organochlorines, was largely responsible for Rachael Carson's book, *Silent Spring*, which precipitated the "environmental crisis" of the 1970s. The organochlorines are generally very persistent broad-spectrum insecticides that tend to be stored in the body fat of organisms. Because of their persistence and massive widespread use, organochlorines have polluted even the snows of Antartica and have accumulated in the fat of all higher carnivores, including *Homo sapiens*. Even now, 12 years after DDT and many other organochlorines were banned in the United States, traces of organochlorines can be detected in human fat, and the eggshells of raptorial birds still collapse sometimes under the weight of brooding parents.

Despite their problems, the organochlorines remain one of the least poisonous of chemicals to humans and other vertebrates, and the effects of DDT on insects, mammals, birds, and fish are better understood than for any other chemical. The demise of the organochlorines resulted, not from their dangerous properties, but from their overuse. Like morphine, DDT is a useful drug, but overdependence on and overuse of both materials can lead to addiction and serious damage.

9.2.2. The Organophosphates

The organophosphate nerve poisons originated in the chemical laboratories of World War II Germany. Because they were intended for use against human targets, organophosphates are generally more toxic to vertebrates than are the organochlorines (Table 9.1). For example, it is estimated that less than two-tenths of a gram of parathion, one of the more potent organophosphates, will kill a human, while more than 30 g of DDT would be required to do the same job. Thus, almost all deaths from insecticide poisoning in agriculture and forestry are caused by organophosphate insecticides.

The organophosphates act on the nervous system by inhibiting acetyl-cholinesterase (AChE), the enzyme that "turns off" nervous impulses after they have been transmitted across a synapse by the transmitter chemical acetylcholine (ACh). The inhibition of AChE causes the synapses to be flooded with the

TABLE 9.1

Characteristics of Some Synthetic Chemical Insecticides Used or Tested Against Forest Insects

Common name	Formula	Half-life (days)	Rat LD$_{50}$ in mg/kg	Toxicity			Used against[a]
				Wildlife	Fish	Bees	
Organochlorines							
HCH	C$_6$H$_6$Cl$_6$	600	91	Low	High	High	Ambrosia beetles, bark beetles, cerambycids
Chlordane	C$_{10}$H$_6$Cl$_8$	300	430	Moderate	High	Moderate	Termites
DDT	C$_{14}$H$_9$Cl$_5$	900	118	Moderate	High	Moderate	Defoliators
Organophosphates							
Acephate	C$_4$H$_{10}$NO$_3$PS	Short	1000	Moderate	Low	High	Budworm, tussock moth, gypsy moth
Azinophos-methyl	C$_{10}$H$_{12}$N$_3$O$_3$PS$_2$	Short	11	Moderate	High	High	Cone and seed insects, aphids, bugs
Carbophenothion	C$_{11}$H$_{16}$ClO$_2$PS$_2$	Short	21	Moderate	?	Moderate	Pine shoot moths
Fenthion	C$_{10}$H$_{15}$O$_3$PS$_2$	<100	200	High	High	High	Sucking insects
Fenitrothion	C$_9$H$_{12}$NO$_5$PS	Short	375	Moderate	High	High	Budworm
Malathion	C$_{10}$H$_{19}$O$_6$PS$_2$	<14	1000	Low	High	High	Budworm
Trichlorfon	C$_4$H$_8$Cl$_3$O$_4$P	Short	500	Moderate	?	Low	Budworm, gypsy moth, tent caterpillar
Carbamates							
Carbaryl	C$_{12}$H$_{11}$NO$_2$	<10	500	Low	Moderate	High	Budworm, gypsy moth, tussock moth, sawflies
Methomyl	C$_5$H$_{10}$N$_2$O$_2$S	Short	20	High	High	High	Budworm
Propoxur	C$_{11}$H$_{15}$NO$_3$	50	90	High	Medium	High	Termites

[a]Mention of the use of a chemical is neither an endorsement nor a recommendation for its use. Insecticide registrations and usage change continuously and should not be employed without the current recommendation of a pest-management specialist.

transmitter chemical so that the nerve endings are continuously stimulated. The poisoned animal dies of exhaustion caused by continuous muscle contraction.

Although many organophosphates are highly toxic, they are usually much less persistent than organochlorines. The half-life of malathion, for example, is less than 2 weeks. Thus, although organophosphate sprays may have a devastating immediate effect, the toxic residues are short-lived and do not accumulate in the food chain.

With the banning of many organochlorine insecticides, organophosphates are being used more and more in forestry, particularly the milder chemicals such as malathion and acephate (Table 9.1). Some organophosphates, such as demetron (Systox), are highly water soluble and can be injected into the sap stream of individual trees to protect them from phloem and meristem feeding insects, e.g., aphids, scales, cone and seed insects, and tip and shoot borers. When used as a systemic insecticide, highly toxic chemicals are less hazardous because fewer organisms come into contact with them.

9.2.3. The Carbamates

A naturally occurring carbamate, eserine, was first isolated in 1864 from a poisonous West African bean plant. It was not until 1947, however, that carbamates were produced commercially for use as insecticides. Carbamates act in much the same way as organophosphates by inhibiting acetylcholinesterase. Carbamates are generally quite toxic to mammals, fish, and useful insects, but have short residual effects (Table 9.1). The most widely used carbamate, carbaryl, has a relatively minor impact on fish and wildlife but can have devastating effects on beneficial parasitic insects and honey bees.

9.2.4. The Botanicals

The botanicals are chemicals such as nicotine, rotenone, and pyrethrin, which occur in plant tissues, presumably as defenses against herbivores. Extracts of these naturally occurring substances have been used for centuries to suppress insect pests, and they are still used today as insecticides and fish poisons. The pyrethrins, in particular, are highly effective insecticides with low mammalian toxicity and short half-lives. They are extracted from the flowers of chrysanthemum plants and are therefore rather expensive. Another problem is that they break down extremely rapidly in the environment. Nowadays, less expensive and more stable synthetic pyrethrins are available, and their use against forest insects may increase in the future, particularly when rapid knockdown of flying insects is required.

9.2.5. The Growth Regulators

Sometimes called the third-generation insecticides, these chemicals either mimic the hormones that control insect development and molting or inhibit the deposition of chitin. Although still in the experimental stage, growth regulators such as Dimilin have been tested successfully against several forest insect pests. Because these chemical act only on the molting and developmental process of arthropods, they usually have little impact on other organisms or on adult insects (e.g., adult parasitoids).

9.2.6. Other Chemicals

Inorganic chemicals such as calcium and lead arsenate were commonly employed as insecticides before the discovery of DDT. Today, they are rarely used against forest insects.

Highly volatile poisons are sometimes used as fumigants. For example, nursery stock and logs may be fumigated prior to export or import as a quarantine procedure, and household pests such as termites and carpenter ants are often controlled by fumigation. Typical fumigants are methyl bromide, ethylene dibromide, and hydrogen cyanide. These volatile chemicals are highly toxic and must be handled with extreme caution.

Petroleum oils are sometimes used to control mites and scale insects on shade trees. The less refined oils are applied as *dormant sprays* during winter, whereas *summer sprays* use lighter more highly refined oils to avoid toxic effects on leaves.

9.2.7. Insecticide Formulation and Application

Insecticides are sold in a variety of forms and must often be diluted or otherwise formulated before application:

Dusts are dry formulations in which the chemical insecticide is diluted in talc, silica gel, clay, or some other inert material. Dusts are usually applied with air-blast machines drawn by tractors; their use is restricted to plantations, seed orchards, and such.

Granular insecticide formulations are similar to dusts except that the particle size is larger. They are applied from hopper-feeder devices, usually to control soil inhabiting insects in nurseries.

Sprays are the most common method of applying insecticides to forests. The chemicals are usually purchased as wettable powders or emulsifiable concentrates and are diluted in water, often with the help of various additives (e.g., wetting agents, emulsifiers, spreaders, adhesives). Insecticides may also be dissolved in organic solvents such as diesel oil. Sprays are usually applied to large forested areas with fixed-wing aircraft or helicopters and to plantations or seed orchards using tractor-drawn or back pack sprayers.

Aerosols are fogs or mists composed of minute insecticidal droplets. The insecticide may be applied in its concentrated form, but in extremely small amounts. Ultralow volume, or ULV, application has become quite popular because of its efficiency in carrying the insecticide to the target organism and the small quantities of material actually applied per unit area of forest. Aerosols are usually applied with air-blast machines that blow the "misted" insecticide with large fans. Helicopters are ideally suited for aerial ULV application. Aerosols must be applied in still air, as they readily drift with the prevailing winds.

Insecticides can also be applied as fumigants to enclosed spaces and as poison baits when incorporated into attractive substances.

9.3. MICROBIAL INSECTICIDES

Insecticide formulations containing pathogenic microorganisms have recently become available commercially. Because of the relative ease of reproducing the pathogen on artificial media, the bacterial entomopathogen *Bacillus thuringiensis,* or *B.t.,* is the most common active ingredient in bio-insecticides. Commercial formulations such as Dipel, Thuricide, and Novabac have been tested successfully against spruce budworms, Douglas-fir tussock moths, gypsy moths, tent caterpillars, and various other forest Lepidoptera.

More recently, virus insecticide formulations have become available. Unlike *B.t.,* viruses have to be grown on their insect hosts and so production is much more expensive. Infected insects are usually collected in the field, or infections are created in laboratory-reared caterpillar populations. The larvae are then freeze-dried, finely ground, and suspended in a water–molasses formulation or with an emulsifiable oil in water suspension. Ultraviolet screening compounds and stickers are often added to prevent the virus from being inactivated by sunlight and to help adhesion to treated foliage.

Because virus particles are spread naturally by contact with infected cadavers or by the movements of parasites and predators, complete spray coverage is not as essential as with chemical insecticides. Viruses are also highly specific, usually infecting only a single species or a few closely related species. Those viruses that have been used as insecticides have not been found to infect other organisms let alone to harm humans, fish, or wildlife, and other useful insects. Because of these characteristics, virus insecticide formulations have great promise for suppressing insect outbreaks. However, production difficulties and costs have impeded commercial development, and the materials are usually stockpiled by government agencies. Formulations currently registered in North America include Gypchek for controlling the gypsy moth, TM Bio Control-1 and Virtuss for the Douglas-fir tussock moth, Lecontvirus for the red-headed pine sawfly, and Neochek-S for the European pine sawfly.

In the future we can look forward to more and different biological insecticides. For example, nematodes have been employed against woodwasps (see Chapter 11) and also show promise against Lepidoptera, Coleoptera (particularly bark beetles; see Chapter 5), and even termites. Possibilities also exist for utilizing fungi and microsporidia in insecticidal sprays.

9.4. BEHAVIORAL CHEMICALS

Chemicals that control the behavior of insects are called behavioral chemicals, behavior-modifying chemicals, or, more technically, *semiochemicals*. These substances may be emitted by an organism to induce behavioral responses in another of the same species, in which case they are called *pheromones*. Alternatively, they may cause behavioral responses in another species, in which case they are called *kairomones* or *allomones* depending on whether they benefit the perceiving or emitting organism, respectively. Some chemicals have more than one function; for example, the aggregation pheromones of certain bark beetles may also act as attractive kairomones for predators of the beetle.

Semiochemicals can also be classified according to the types of behavior they elicit in the responding organism. For instance, *attractants* cause the organism to orientate toward an odor source, *repellents* have the opposite effect, *incitants* induce a particular behavior such as feeding, and *deterrents* cause the animal to refrain from that behavior. Attractants, repellents, and deterrents are discussed below because of their potential for suppressing pest populations.

9.4.1. Chemical Attractants

Three kinds of attractants show promise for forest pest population suppression:

1. *Kairomones emitted by host plants:* Most herbivorous insects orient toward chemical odors emanating from plants of the species on which they feed, or from particular individual plants within the host population (i.e., dead, weak, or unhealthy trees). Predators and parasitoids are also attracted by chemicals emitted by their insect prey and/or by the plants on which their prey feed.
2. *Sex pheromones:* These attract one sex, usually the male, for the purpose of mating. Sex pheromones are probably produced by most insects but their development as population suppressants is more advanced in the Lepidoptera (moths and butterflies).
3. *Aggregation pheromones:* These attract both sexes to favorable habitats such as suitable food sources, reproductive substrates, hibernation sites, nests, or hives.

In general, the effectiveness of a particular attractant depends on its power of attraction (attractiveness) and the proportion of the population that responds to the chemical. Aggregation pheromones are particularly effective because they are often very powerful and attract both sexes. Sex pheromones are also highly attractive but only influence the behavior of one sex, whereas host kairomones are usually less powerful and may or may not be sex selective. Nevertheless, attractants of all types should be considered as potential pest population suppressants.

There are two basic approaches for using attractants to control pest popula-

tions—*trapping* and *behavior disruption*. In the first method, the chemical attractant is used to guide insects to a point source where they are trapped or
otherwise destroyed. This approach has been used for centuries to suppress bark
beetle populations in Europe. In the early days *trap trees* were employed to
capture and eliminate *Ips* beetles. Several host trees were usually felled for each
infested tree, and beetles attacking the trap trees were eliminated by burning or
insecticidal treatment. This method uses both host-produced kairomones and the
insect's aggregation pheromones to concentrate the population. With the recent
identification and synthesis of aggregation pheromones for many important bark
beetles, other trapping methods have evolved; e.g., pheromone-baited sticky
traps, pipe and funnel traps that simulate trees, window traps, and pheromone-
baited trap trees (Fig. 9.5). Some of these semiochemical formulations are now
available commercially and have been used experimentally to suppress bark
beetle populations (Table 9.2). Pheromone-baited pipe traps were employed, for
example, in an effort to suppress a large spruce bark beetle outbreak in Scandinavia (Table 9.2). More than half a million traps were deployed in southern
Norway alone, and more than seven billion beetles were trapped in 1979 and
1980. Even though this massive trapping effort probably had a significant impact

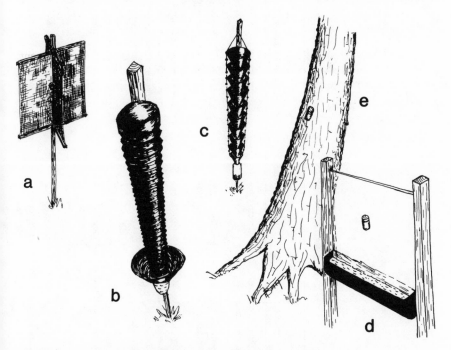

FIGURE 9.5. Styles of traps used in conjunction with pheromone–kairomone formulations to trap
bark beetles: (a) sticky vane trap; (b) pipe trap; (c) multiple funnel trap; (d) window vane trap; (e) trap
tree baited with pheromone and treated with insecticide or herbicide.

TABLE 9.2
Some Attractants Tested or Used Operationally to Suppress Forest Pest Populations

Insect	Chemical name (trade name)	Chemical origin and formulation	Objective/success	Reference
Western pine-shoot borer *Eucosma sonomana*	Dodecenyl acetate	Sex pheromone in ConRel fibers or Hercon flakes sprayed from aircraft	Mating disruption 70–80% suppression	Daterman (1982)
Gypsy moth *Lymantria dispar*	Expoxy methyloctadecane (Disparlure)	Sex pheromone in microcapsules or ConRel fibers sprayed by aircraft	Mating disruption 80–96% suppression	Cameron (1981)
Douglas-fir tussock moth *Orgyia pseudotsugata*	Heneicosenone	Sex pheromone in ConRel fibers sprayed from aircraft	Mating disruption >77% suppression	Sower et al. (1979)
Spruce bark beetle *Ips typographus*	Methylbutenol cis-Verbenol Ipsdienol	Aggregation pheromones Plastic strip or bag in pipe traps	Mass trapping	Bakke et al. (1983)
Western pine beetle *Dendroctonus brevicomis*	exo-Brevicomin Frontalin Myrcene	Aggregation pheromone Aggregation pheromone Host kairomone Sticky traps	Mass trapping Reduced tree mortality by 68%	Wood (1979)
Mountain pine beetle *D. ponderosae* in western white pine	trans-Verbenol α-Pinene (Pondelure)	Aggregation pheromone Host kairomone	Mass trapping	Pitman (1971)

in lodgepole pine	trans-Verbenol exo-Brevicomin Myrcene	Aggregation pheromone Aggregation pheromone Host kairomone Baited trees	Containment and concentration in trap trees	Borden et al. (1983)
Southern pine beetle D. frontalis	Frontalin α-Pinene (Frontalure)	Aggregation pheromone Host kairomone Baited trees	Trap trees treated with cacodylic acid Behavior disruption	Billings (1982)
Ambrosia beetle Gnathotrichus sulcatus	Sucatol 2:1 mixture of S:R enantiomers	Aggregation pheromone in vial dispensers	Mass trapping >60% suppression	Borden and McLean (1981)
	Ethanol α-Pinene	Aggregation pheromones Host kairomone Host kairomone Sticky vane traps and multiple funnel traps	Mass trapping	Lindgren and Borden (1983)
Ambrosia beetle Trypodendron lineatum	Lineatin Ethanol α-Pinene	Aggregation pheromone Host kairomone Host kairomone Sticky vane traps and multiple funnel traps	Mass trapping 77% suppression	Lindgren and Borden (1983)
Smaller European elm beetle Scolytus multistriatus	α-Multistriatin Methylheptanol α-Cubebene (Multilure)	Aggregation pheromone Aggregation pheromone Host kairomone ConRel fibers or Hercon strips and sticky vane traps	Mass trapping reduced rate of Dutch elm disease spread in some tests	Lanier (1981) Peacock et al. (1981)

on the beetle population, it is not clear whether the subsequent decline in spruce mortality was caused by the suppression operation or by natural causes.

Pheromone trapping systems have also been used on an operational scale to suppress ambrosia beetle populations in timber processing areas (Table 9.2). In this situation, the treatment area is relatively small and the objective is to prevent beetle attacks on logs awaiting processing. Although large numbers of beetles can be caught in sticky traps or funnel traps, the actual reduction in log damage is again difficult to determine.

Pheromones are also employed in a modification of the old trap-tree approach. In this case, trees that are the most suitable for beetle reproduction are baited with attractive pheromones and the trees are subsequently logged and the beetles killed in the milling process. Alternatively, baits are placed on hosts that have been treated with insecticides or herbicides (Table 9.2). Herbicide-killed trees usually form a poor environment for bark beetle development, and most of the brood die before emergence.

The second way in which attractants can be used to suppress pest populations is to disrupt or confuse the normal behavior of the insect. Organisms "home in" on an attractive chemical by following an *odor plume*. The insect locates the source of this odor plume by flying upwind to its origin. The more unique the chemical, the sharper the odor plume boundary because there is less background interference. This is why pheromones, unique to the species, are such powerful attractants, whereas kairomones are often found in a number of different species; e.g., α-pinene and myrcene, host kairomones for the mountain pine beetle and Douglas-fir beetle, are emitted by both host and nonhost conifers.

The idea behind behavior disruption is to increase the background interference to a level at which the insect has difficulty locating the correct odor source. This effect can be achieved by saturing the atmosphere with the attractive material or by creating many artificial odor sources in the environment. In the former the insect has difficulty detecting an odor plume, while in the latter it wastes time and energy after the wrong odor trails. Technologically, artificial odor sources seem to be the most feasible approach to disrupt insect behavior. Pheromones can be incorporated into various plastic materials from which they are released at a controlled rate over a period of several weeks. Small plastic strips or microcapsules containing attractive pheromones and coated with sticky material can then be sprayed on trees with ground or air applicators (Table 9.2).

9.4.2. Chemical Repellents and Deterrents

Most chemical repellents and deterrents are plant-produced allomones that inhibit insect attack or feeding (Table 9.3). Some bark beetles, however, are known to emit so-called antiaggregation pheromones that repel members of the same species from trees that have been successfully colonized. The natural function of antiaggregation pheromones is to prevent overpopulation of infested

TABLE 9.3
Chemicals That Have Repellent or Deterrent Activity against Forest Insects

Insect	Chemical (action)	Chemical origin and formulation	Objective/success	Reference
Mountain pine beetle *Dendroctonus ponderosae*	Pine oil (repellent or deterrent)	Pulp residue (host) sprayed on trees	Prevented attack successful	Nijholt *et al.* (1981)
Douglas-fir beetle *D. pseudotsugae*	MCH (repellent)	Antiaggregation pheromone in plastic beads dispersed by helicopter	Reduced attack on wind-throws by >90%	Furniss *et al.* (1981)
Southern pine beetle *D. frontalis*	*endo*-Brevicomin *exo*-Brevicomin Verbenone	Antiaggregation pheromones released on individual trees	Reduced gallery construction by 92%, eggs by 84%	Payne and Richerson (1979)
Smaller European elm beetle *Scolytus multistriatus*	Juglone (deterrent)	Nonhost allomone	Not tested operationally	Norris (1970)
Fir engraver beetle *S. ventralis*	Limonene Δ^3-Carene α-Pinene Myrcene β-Pinene (repellents)	Host allomone	Not tested operationally	Bordasch and Berryman (1977)
Ambrosia beetles *Trypodendron* and others	Pine oil (repellent or deterrent)	Pulp residue (host) sprayed on logs	Retarded and reduced attacks	Nijholt (1980)
White pine weevil *Pissodes strobi*	Pine oil (deterrent)	Pulp residue (host) sprayed on trees	Inhibits feeding in laboratory tests	Alfaro *et al.* (1984)
Pine cone weevil *Pissodes validirostris*	α-Pinene (repellent or deterrent)	Host allomone	Not tested operationally	Annila and Hiltunen (1977)
Jack pine sawfly *Neodiprion swainii*	Ketopodocarpenoic acid (deterrent)	Host allomone	Laboratory test Inhibited feeding	Ikeda *et al.* (1977)
European pine sawfly *N. sertifer*	Same	Sprayed on foliage	Inhibited feeding	Niemelä *et al.* (1982)

trees and thereby to minimize mortality from intraspecific competition. Antiaggregation pheromones have been successfully employed to suppress bark beetle attacks on susceptible host material, such as windthrown trees (Table 9.3). Theoretically they could also be used to protect individual trees or areas or to confuse the insect's orientation to its host.

Although numerous plant chemicals have been identified as repellents and feeding deterrents, few attempts have been made to test them on an operational basis. The exception is pine oil, a by-product of the paper industry, which has been successfully tested against several forest pests (Table 9.3). In general, repellents and deterrents show most promise for protecting individual trees of exceptional value, such as shade trees in urban areas, parks, and campgrounds.

9.5. GENETIC MANIPULATION

The inception of genetic methods of insect control was the so-called *sterile male technique* wherein male insects were sterilized and then released into the wild population. If sufficient sterile males can be released to reduce the birth rate below the death rate, the population will automatically decline (i.e., $b < d$; see Chapter 4). In theory the continued release of sufficient sterile males will eventually cause the population to decline to *extinction*. Although this technique has been employed against several agricultural pests, technical and logistical problems are often insurmountable in forest environments. For example, the success of the method depends on several factors:

1. A method for rearing huge numbers of insects.
2. The sterilizing treatment, usually irradiation with gamma rays or treatment with chemosterilants, must not affect the dispersal or mating ability of the sterile males.
3. The sterile males must be distributed or disperse themselves throughout the wild population in proportion to the density of that population.
4. Sufficient sterile males must be released to reduce the birth rate below the death rate.
5. Females must mate only once or, if they mate more than once, the sterile spermatozoa must be equally likely to fertilize an egg.

A number of other ideas, such as producing sterile hybrids or introducing lethal genetic strains, have been proposed but remain largely untested. In the future, however, and with the growth of *genetic engineering,* we can expect to hear more about genetic methods of pest control.

9.6. PHYSICAL AND MECHANICAL METHODS

Before the advent of cheap insecticides, physical and mechanical methods were often used for insect population suppression. For example, bark beetle populations were often attacked with the *fell–peel–burn* method, in which infested trees were cut and the bark peeled and burned. In warmer regions, and on south-facing slopes, beetle-infested trees were often sun cured by exposing them to direct sunlight. In this case logs had to be rolled periodically so that all beetles were exposed to the high temperatures generated by direct exposure to the sun. The removal and disposal of trees infested by bark beetles still has a place in modern forestry, particularly if the costs can be reduced by utilizing the infested material. For example, control logging was used against the huge Engelmann spruce beetle outbreak that occurred in northern Idaho and Montana from 1952 to 1955. Beetle-killed spruce were logged and milled as rapidly as possible in an attempt to salvage merchantable timber and, at the same time, to reduce the bark beetle population. Single-tree disposal, in which individual trees are cut, bunched, piled, and burned, is also routinely used to suppress mountain pine beetle infestations in western Canada. The part played by control logging in the decline of bark beetle outbreaks has not been proved, however.

Water treatment is still a practical and efficient method for killing beetles within infested trees or for preventing attacks on fresh timber or lumber. Submersion of logs in holding ponds or spraying log decks with water are effective in reducing beetle attacks. The manager should be aware, however, that ambrosia beetle attacks often occur on unsubmerged or unwatered log sections.

Other physical and mechanical methods that have been tried with varying degrees of success include blasting the bark off beetle infested trees by wrapping them with primer chord, placing sticky bands around tree trunks to trap migrating defoliator larvae, and using light traps to capture night-flying moths.

9.7. FOOD REMOVAL

Manipulating food supplies, or host plants, is usually considered a preventive measure (Chapter 7). However, this approach can also be used for containment or suppression. For instance, food removal is sometimes used to eliminate small local infestations that have penetrated containment barriers. In such cases all the host trees in, and for a certain distance around, the infested area are usually destroyed. Food removal has also been used to suppress bark beetle populations. For example, the growth of southern pine beetle "hot spots" can be retarded or even halted by felling a buffer strip along the leading edge of the infestation, while mountain pine beetle populations can be reduced by removing the large-diameter trees, or those trees with thick phloem, which are most suitable for the development of beetle larvae.

As we have seen, innumerable tactics have been suggested or tried for controlling forest pest outbreaks, and we have not covered them all by far. In the end, control tactics are only limited by the imagination and ingenuity of the inventor. We will undoubtedly hear many new ideas from the next generation of forest entomologists.

REFERENCES AND SELECTED READINGS

Alfaro, R. I., Borden, J. H., Harris, L. H., Nijholt, W. W., and McMullen, L. H., 1984, Pine oil, a
feeding deterrent for the white pine weevil, *Pissodes strobi* (Coleoptera: Curculionidae), *Can.
Entomol.* **116:**41–44.

Annila, E., and Hiltunen, R., 1977, Damage by *Pissodes validirostris* (Coleoptera, Curculionidae)
studied in relation to the monoterpene composition in Scots pine and lodgepole pine, *Ann.
Entomol. Fenn.* **43:**87–92. (repellent or deterrant).

Auer, C., Roques, A., Goussard, F., and Charles, P.-J., 1981, Effets de l'accroissement provoqué du
niveau de population de la tordeuse du mélèze *Zeiraphera diniana* Guéneé (Lep., Tortricidae)
au cours de la phase de régression dans une massif forestière du Brianconnais, *Z. Angew.
Entomol.* **92:**286–303. (suppression of population cycles)

Bakke, A., 1981, The utilization of aggregation pheromone for the control of the spruce bark beetle,
in: *Insect Pheromone Technology: Chemistry and Applications* (B. A. Leonhardt and M.
Beroza, eds.), pp. 219–229, American Chemical Society Symposium Series No. 190, ACS.

Bakke, A., Saether, T., and Kvamme, T., 1983, Mass trapping of the spruce bark beetle *Ips
typographus:* Pheromone and trap technology, *Rep. Norwegian For. Res. Inst.* **38**(3):1–35.

Berryman, A. A., Amman, G. D., Stark, R. W., and Kibbee, D. L., 1978, Theory and practice of
mountain pine beetle management in lodgepole pine forests, Forest, Wildlife and Range Experiment Station, University of Idaho, Moscow. (see papers by W. H. Klein on chemical and
mechanical tactics, H. S. Whitney, L. Safranyik, S. J. Muraro, and E. D. A. Dyer on ex-
plosives and burning, G. B. Pitman, M. W. Stock, and R. C. McKnight on pheromones, D. B.
Cahill and D. R. Hamel on cutting strategies)

Billings, R. F., 1982, Direct control, in: The Southern Pine Beetle (R. C. Thatcher, J. L. Searcy, J.
E. Coster, and G. D. Hertel, eds.), pp. 179–192, U.S. Forest Service Technical Bulletin 1631.
(suppressing "hot spots" with buffer strips)

Bordasch, R. P., and Berryman, A. A., 1977, Host resistance to the fir engraver beetle, *Scolytus
ventralis* (Coleoptera: Scolytidae). 2. Repellency of *Abies grandis* resins and some monoterpenes, *Can. Entomol.* **109:**95–100.

Borden, J. H., 1971, Changing philosophy in forest–insect management, *Bull. Entomol. Soc. Am.*
17:268–273. (examples of secondary pest outbreaks and a selective, nonpersistent insecticides)

Borden, J. H., 1982, Aggregation pheromones, in: *Bark Beetles in North American Conifers: A
System for the Study of Evolutionary Biology* (J. B. Mitton and K. B. Sturgeon, eds.), pp. 74–
139, University of Texas Press, Austin.

Borden, J. H., Chong, L. J., and Fuchs, M. C., 1983, Application of semiochemicals in post-
logging manipulation of the mountain pine beetle, *Dendroctonus ponderosae* (Coleoptera:
Scolytidae), *J. Econ. Entomol.* **76:**1428–1432.

Borden, J. H., Chong, J. L., Pratt, K. E. G., and Gray, D. R., 1983, The application of behavior-
modifying chemicals to contain infestations of the mountain pine beetle, *Dendroctonus ponderosae, For. Chron.* **59:**235–239.

Borden, J. H., and McLean, J. A., 1981, Pheromone-based suppression of ambrosia beetles in
industrial timber processing areas, in: *Management of Insect Pests with Semiochemicals* (E. R.
Mitchell, ed.), pp. 133–154, Plenum, New York.

Burgess, H. D. (ed.), 1981, *Microbial Control of Pests and Plant Diseases 1970–1980,* Academic
Press, New York.

Cameron, E. A. (and others), 1981, The use of disparlure to disrupt mating, in: The Gypsy Moth:
Research towards Integrated Pest Management (C. C. Doane and M. L. McManus, eds.), pp.
554–572, U.S. Forest Service Technical Bulletin 1584.

Cole, W. E., and McGregor, M. D., 1985, Reducing or preventing mountain pine beetle outbreaks in
lodgepole pine stands by selective cutting, in: The Role of the Host in the Population Dynamics
of Forest Insects, Proceedings of the IUFRO Conference, Banff, Canada (L. Safranyik, ed.),
pp. 175–185, Pacific Forest Research Centre, Victoria, British Columbia.

Cunningham, J. C., Tonks, N. V., and Kaupp, W. J., 1981, Viruses to control winter moth, *Operophtera brumata* (Lepidoptera: Geometridae), *J. Entomol. Soc. BC* **78**:17–24.

Daterman, G. E., 1982, Control of western pine shoot borer damage by mating disruption—A reality, in: *Insect Suppression with Controlled Release Pheromone Systems* (A. F. Kydonieus, M. Beroza, and G. Zweig, eds.), pp. 155–163, CRC Press, Boca Raton, Florida.

Fellin, D. G., 1983, Chemical insecticide vs the western spruce budworm: After three decades, what's the score?, *West. Wildl.* **9**:8–12.

Furniss, M. M., Clausen, R. W., Markin, G. P., McGregor, M. D., and Livingston, R. L., 1981, Effectiveness of Douglas-fir beetle antiaggregative pheromone applied by helicopter, U.S. Forest Service General Technical Report INT-101.

Hamel, D. R., 1981, Forest-management chemicals: A guide to use when considering pesticides for forest management, U.S. Forest Service Agriculture Handbook 585.

Harper, J. D., 1974, Forest Insect Control with *Bacillus thuringiensis:* Survey of Current Knowledge, University Printing Service, Auburn University, Auburn.

Ikeda, T., Matsumura, F., and Benjamin, D. M., 1977, Mechanism of feeding discrimination between matured and juvenile foliage by two species of pine sawflies, *J. Chem. Ecol.* **3**:677–694. (deterrents)

Kurstak, E. (ed.), 1982, *Microbial and Viral Pesticides,* Marcel Dekker, New York.

Kydonieus, A. F., Beroza, M., and Zweig, G., 1982, *Insect Suppression with Controlled Release Pheromone Systems,* Vols. 1 and 2, CRC Press, Boca Raton, Florida.

Lanier, G. N., 1981, Pheromone-baited traps and trap trees in the integrated management of bark beetles in urban areas, in: *Management of Insect Pests with Semiochemicals* (E. R. Mitchell (ed.), pp. 115–131, Plenum, New York.

Leonhardt, B. A., and Beroza, M. (eds.), 1982, *Insect Pheromone Technology: Chemistry and Applications,* American Chemical Society Symposium Series No. 190.

Lewis, F. B., McManus, M. L., and Schneeberger, N. F., 1979, Guidelines for the use of GYPCHEK to control the gypsy moth, U.S. Forest Service Research Paper NE-441. (virus insecticide)

Lindgren, S., and Borden, J. H., 1983, Survey and mass trapping of ambrosia beetles (Coleoptera: Scolytidae) in timber processing areas on Vancouver Island, *Can. J. For. Res.* **13**:481–493.

Martouret, D., and Auer, C., 1977, Effets de *Bacillus thuringiensis* chez une population de tordeuse grise du Mélèze, *Zeiraphera diniana* (Lep.: Tortricidae) en culmination gradologigue, *Entomophaga* **22**:37–44.

McGregor, M. D., Furniss, M. M., Oakes, R. D., Gibson, R. E., and Meyer, H. E., 1984, MCH pheromone for preventing Douglas-fir beetle infestation in windthrown trees, *J. For.* **82**:613–615.

McLean, J. A., and Borden, J. H., 1977, Suppression of *Gnathotrichus sulcatus* with sucatol-baited traps in a commercial sawmill and notes on the occurrence of *G. retusus* and *Trypodendron lineatum, Can. J. For. Res.* **7**:348–356.

Mitchell, E. R. (ed.), 1981, *Management of Insect Pests with Semiochemicals,* Plenum Press, New York.

Morris, O. N., 1980, Entomopathogenic viruses: Strategies for use in forest insect pest management, *Can. Entomol.* **112**:573–584.

Morris, O. N., Dimond, J. B., and Lewis, F. B., 1984, Guidelines for the operational use of *Bacillus thuringiensis* against the spruce budworm, U.S. Forest Service Agriculture Handbook No. 621.

National Academy of Sciences, 1972, Pest Control Strategies for the Future, National Academy of Sciences (USDA), Washington, D.C.

Niemelä, P., Mannila, R., and Mantasla, P., 1982, Deterrent in Scots pine, *Pinus sylvestris,* influencing feeding behavior of the larvae of *Neodiprion sertifer* (Hymenoptera, Diprionidae), *Ann. Entomol. Fenn.* **48**:57–59.

Nijholt, W. W., 1980, Pine oil and oleic acid delay and reduce attacks on logs by ambrosia beetles (Coleoptera: Scolytidae), *Can. Entomol.* **112**:199–204.

Nijholt, W. W., McMullen, L. H., and Safranyik, L., 1981, Pine oil protects living trees from attack by three bark beetle species, *Dendroctonus* spp., *Can. Entomol.* **113:**337–340.

Norris, D. M., 1970, Quinol stimulation and quinone deterrency of gustation by *Scolytus multistriatus, Ann. Entomol. Soc. Am.* **63:**476–478.

Payne, T. L., and Richerson, J. V., 1979, Management implications of inhibitors for *Dendroctonus frontalis* (Col., Scolytidae), *Bull. Soc. Entomol. Suisse* **52:**323–331.

Peacock, J. W., Cuthbert, R. A., and Lanier, G. N., 1981, Deployment of traps in a barrier strategy to reduce populations of the European elm bark beetle, and the incidence of Dutch elm disease, in: *Management of Insect Pests with Semiochemicals* (E. R. Mitchell, ed.), pp. 155–174, Plenum, New York.

Prebble, M. L. (ed.), 1975, Aerial Control of Forest Insects in Canada, Department of the Environment, Ottawa.

Retnakaran, A., Grant, G. G., Ennis, T. J., Fast, P. G., Arif, B. M., Tyrrell, D., and Wilson, G., 1982, Development of environmentally acceptable methods for controlling insect pests of forests, Canadian Forestry Service Information Report FRM-X-62.

Ritter, F. T. (ed.), 1979, *Chemical Ecology: About Communication in Animals,* Elsevier/North-Holland Biomedical Press, New York.

Smith, R. H., 1976, Low concentration of lindane plus induced attraction traps mountain pine beetle, U.S. Forest Service Research Note PSW-316.

Sower, L. L., Daterman, G. E., Orchard, R. D., and Sartwell, C., 1979, Reduction of Douglas-fir tussock moth reproduction with synthetic pheromone, *J. Econ. Entomol.* **72:**739–742.

Wood, D. L., 1979, Development of behavior modifying chemicals for use in forest pest management, in: *Chemical Ecology: About Communication in Animals* (F. T. Ritter, ed.), pp. 261–279, Elsevier/North-Holland Biomedical Press, New York.

CHAPTER 10

PEST MANAGEMENT DECISIONS

Management is the *art* of manipulating men, money, machines, and resources in order to attain certain predefined goals. Although management is essentially an art, like other art forms it should rest on a firm foundation of scientific knowledge. Architecture is also an art, but our bridges and towers would not stand long if the architect's plans ignored physical laws and material properties. Thus, the failure of management planning is often due to ignorance or misinterpretation of the "laws" of nature rather than to ineffective application. The successful manager is an applied scientist who uses all the relevant results of scientific research to achieve his or her goals.

The forest manager is charged with the maintenance of forest resources for the benefit of public or private owners. As part of this task, he or she has to consider biological, social, economic, and political realities, some of which may be in serious conflict. Because forests are biological entities, however, it is essential that the manager have a sound scientific understanding of the biological system being managed. Part of this biological system is the complex of insects feeding on forest trees, pollinating them, and feeding on other insects.

The protection of forests from damage by insects is one of the many concerns of the forest manager. Consideration must also be given to other destructive agents, such as fire, diseases, and wind, as well as harvesting, planting, thinning, and other forestry operations. The major problem the manager faces is setting priorities and deciding how much of a limited money supply to channel into these competing management activities.

In order to make informed decisions concerning insect pest management operations, the forest manager needs answers to the following questions:

1. Where on my forested lands is damage currently occurring, how much damage is being done, and will this damage increase or decrease in the

near future? These concerns about the present status of pest damage, and the probable short-term course of infestations, are addressed by monitoring the forest and making short-term forecasts (Chapter 6).

2. Given that I have a serious insect problem on my lands, what can I do to reduce or eliminate the damage? Various containment or suppression tactics provide the manager with a number of alternative courses of action (Chapter 9).

3. Where on my forested lands is damage likely to occur in the future? These concerns with conditions in the more distant future are addressed by risk assessment methods (Chapter 7).

4. What can I do to reduce the likelihood of damage in the future? The problem of preventing insect-caused damage can be approached by silvicultural and natural enemy manipulations (Chapter 8).

The preceding chapters in this book deal with techniques for addressing the major protection questions raised by forest managers. They do not, however, deal with methods for deciding *which* alternatives to utilize in a given situation. For instance, the manager might also ask the following questions:

5. Given that an insect outbreak has been detected on my lands, what single treatment or combination of treatments should I use to contain and/or suppress the infestation—or should I do nothing?

6. How can I organize or change my current management operations in order to reduce the chances of insect outbreaks later on?

7. Should I wait for a destructive outbreak to occur and then try to suppress it or should I try and prevent its occurrence in the first place?

These questions, and others like them, require the forest manager to make decisions.

Decisions are made in all walks of life by comparing various alternatives and then choosing the one that best suits some predetermined objective(s). In order to make informed decisions, we must first specify our objective(s) or define what economists call the *objective function*. For example, if the forest were being managed for commercial timber production, the objective function might be to maximize net wood production or net profits over time. If, on the other hand, multiple forest uses were being considered, the objective function might be to maximize a number of different forest values over time.

In order to give fair weight to several different forest uses, they need to be converted into some common denominator. In many societies the common value system is money, while others may also consider the more subjective criterion of social good. Using money as the common denominator poses some problems for societies that value the esthetic properties of nature. How does one place a monetary value on a panoramic vista, a moose, a woodpecker? If a fair decision is to be made, however, an attempt must be made to reduce these intangibles to the common currency. Alternatively, the intangibles can be treated as constraints

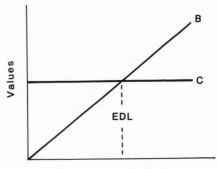

FIGURE 10.1. The economic damage level (EDL) is specified as that insect population density that causes damage in excess of the costs of control or as the point at which the benefits of control (B), in terms of damage saved, equal the costs of control (C).

on management decisions. In other words, the use of particular options is constrained or even prohibited if they cause serious impacts to valued resources. Thus, DDT was banned in the United States because of its unacceptable damage to natural ecosystems.

In Chapter 9 we introduced the idea of an *economic damage level* (EDL), where decisions to act against an insect outbreak were made if the economic losses are expected to exceed the costs of control. This idea is illustrated in Fig. 10.1. In its simplest sense, the decision to take action against an insect pest rests on our ability to estimate the expected economic impact of the outbreak and the costs that will be incurred in controlling it. Alternatively, we may wish to compare a number of management actions by evaluating how effectively they reduce the damage as compared to their costs; that is,

$$\text{Net benefit of action} = \text{value of resources saved} - \text{cost of action} \quad (10.1)$$

Irrespective of the type of decision being made, the manager requires information on the expected impacts of insect infestations on all resources under consideration as well as on the costs of the options being considered.

10.1. ESTIMATING ECONOMIC IMPACTS

Insect infestations can cause direct impacts on forest values by killing large trees, seedlings, and seeds, by reducing diameter and height growth, and by deforming the stem. These impacts, particularly tree mortality, can also have indirect effects on other resources such as wildlife, fish, water, and recreation.

10.1.1. Mortality of Mature Trees

Certainly the most spectacular and economically significant impact of insect outbreaks is the mortality they cause in mature forests. Large, mature trees usually have a high market and esthetic value and their replacement requires long-term investments in man-power and money.

Because insect-killed timber has salvage value, however, the total loss should be adjusted by subtracting the expected gains from salvage operations. To be effective, salvage must be carried out at the earliest possible time because dead trees quickly deteriorate from woodboring insects (roundheaded and flatheaded borers, ambrosia beetles, and woodwasps) and decay-causing micro-organisms. These decay agents reduce the volume of salvageable timber, lower its value, and increase processing costs.

The rate of deterioration of dead trees is affected by a number of variables, particularly the size and species of tree, the cause of death, and the prevailing environmental conditions. For instance, such species as balsam fir and hemlock often deteriorate completely in as few as 3 years, while other species such as cedar and cypress may yield saleable lumber after several decades. Old-growth, large-diameter trees also tend to deteriorate more slowly than small young ones, and deterioration usually proceeds more slowly in cool, dry environments.

The deterioration of dead timber can be estimated by sampling felled trees in logging areas and calculating volume lost from the cross-sectional area destroyed by insects and decay. Pulpwood losses can be assessed by measuring the average diameter, depth, and density of the insect borings and converting this to volume reduction. Calculating the monetary value of insect-infested timber may require additional information, however. For example, infested timber may have lower grades or market demand, and costs may be incurred in processing the lumber; i.e., sawing patterns and sorting procedures may have to be changed. On the other hand, timber values may sometimes be increased by insect infestation, as in the speciality market for ''blue-stained'' and engraved or ''wormy'' wood (particularly the furniture industry).

Hidden costs can also occur if accumulations of insect-killed timber saturate the market, causing depressed prices for green as well as infested lumber, and if insect infestations interfere with even-flow, sustained yield plans. In addition, costs may be incurred if dead trees pose a fire hazard and if infested areas have to be replanted.

Besides timber values, tree mortality can also influence other forest resources (Fig. 10.2). For example, populations of large herbivorous mammals such as deer and elk probably benefit because openings in the canopy stimulate increased forage production, and fallen trees may provide protective bedding places. Similarly, we may find an increase in cattle forage and an economic benefit in terms of the number of grazing permits that can be issued. In extreme cases, however, where large numbers of fallen trees accumulate, the forest may become impenetrable for cattle and big game.

Recreational use of the forest may also be impacted, particularly when mortality occurs in campgrounds and scenic vistas. Most studies, however, support the view that recreation is not seriously affected by insect infestations,

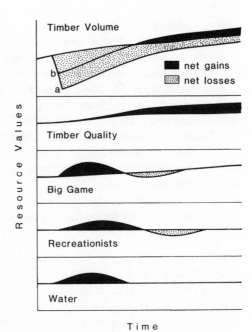

FIGURE 10.2. Hypothetical impact of tree mortality due to a insect outbreak on the values of forest resources over time. (Top) Timber volume losses can be reduced by salvage; a, losses without salvage; b, losses when 40% of the volume is salvaged.

because recreationists have specific goals, such as hunting, fishing, and climbing. On the other hand, the impediment to movement imposed by fallen trees, not to mention the danger and possible lawsuits, may cause serious problems in later years.

Water production may be increased by insect infestations, sometimes for many decades (Fig. 10.2). This results because plant cover is removed, decreasing evapotranspiration and increasing spring runoff from unshaded snow. Increased streamflow, however, can cause negative impacts on fisheries by scouring spawning beds and changing the habitat for aquatic insects which form the major food of many game fish. Erosion can also be a problem in certain areas, particularly steep slopes with shallow duff layers and sparse ground vegetation. However, accumulations of dead leaves, twigs, and so forth, and the rapid growth of the understory normally prevent serious erosion.

Calculating the monetary impacts of insect infestations on diverse forest resources is rarely a simple matter. Recreational values can be assessed by monitoring visitor-days and then using standard dollar-value per visitor-day estimates. The value of increased forage for cattle and big game production can be obtained by estimating the additional numbers of game and cattle that can be supported and multiplying by the assumed dollar value of each animal.

The economic impact of timber mortality can change considerably with the date of final harvest. Timber losses may be small if the stand is harvested soon after the outbreak or if it is not scheduled for harvest for a long time (Fig. 10.2); in the latter case, accelerated growth rates may compensate for much of the loss and there may even be a net gain if salvage was possible.

Lumber quality should also be considered when calculating long-term damage expectations. In some cases thinning the smaller trees and recycling nutrients to the more vigorous stems may lead to stands with fewer but higher-quality trees (Fig. 10.2).

Finally, one should realize that income from timber sales only occurs infrequently during a rotation (e.g., commercial thinnings and harvest), whereas income from grazing and hunting permits, water production, recreation, and such accrues annually.

10.1.2. Mortality of Immature Trees and Seed

Mortality of seeds and seedlings has a major impact on reforestation efforts, slowing the establishment of preferred tree species or necessitating artificial regeneration or repeated replanting. Besides the additional costs of regeneration, losses may also be incurred because harvesting is delayed or because less valuable species occupy the site. Mortality to seed and seedlings can be caused by cone and seed insects, tip-boring insects, defoliators, and root insects. Cone and seed insects are particularly important in naturally regenerating areas during poor cone years, whereas tip and shoot insects and root weevils may completely destroy natural or planted seedlings. Defoliators may cause problems by feeding on flowers and cones and by descending to feed on the understory trees when the overstory has been heavily defoliated.

The impact of insects on forest regeneration can be assessed by estimating replanting costs and discounting timber values due to delayed harvest. Impacts on other resources may also have to be estimated when forest conditions are seriously altered by the infestation.

10.1.3. Growth Loss

Trees which survive insect attack frequently suffer growth reduction, particularly when their photosynthesizing leaf area has been reduced by defoliation, nutrients removed from their phloem by sap sucking insects, or if their apical leaders have been killed by tip and shoot insects. In mature trees, which have reached their maximum height, the main losses occur in diameter increment. Reductions in radial increment can be estimated from a sample of trees by taking increment cores and measuring the width of the growth rings (Fig. 7.2). Growth impact is then determined by comparison with a sample of nearby unattacked trees or, if this is not possible, by measuring the width of growth rings prior to defoliation. These sample statistics can then be converted to volume reductions over the sampling universe to provide an estimate of growth loss for a specified forest area.

When estimating the impact of insect defoliation on radial growth, it should be realized that trees can recover very rapidly after the insect population declines. Defoliated trees, in fact, may then grow faster than their undefoliated neighbors so that volume losses may be recovered rapidly following defoliation (Fig. 10.3). This growth acceleration is probably due to stand thinning and increased soil nutrients released from insect feces and decaying bodies (Chapter 3). Ideally, estimates of growth impact should be adjusted to reflect the expected growth losses at harvest. Thus, growth loss estimates, which may be quite high if trees are harvested during or immediately after the outbreak, should be reduced according to the expected growth recovery when trees are harvested in the following years.

Height growth reductions due to infestations by tip and shoot insects, defoliators, sap-suckers, and other insects may impact the development of trees and stands. Height increment losses to seedlings, saplings, and pole size trees, for instance, can reduce their competitive edge relative to other trees and shrubs. This may result in undesirable species occupying the site or in delayed harvest or lower volumes at harvest. Volume losses due to height-growth reductions can be estimated by projecting tree taper curves for infested and uninfested trees over the rotation period. However, effects on the competitive status of infested trees are much more difficult to estimate and may be much more serious. If height growth is reduced to such an extent that brush or other noncommercial species take over the site, heavy costs may be incurred for site preparation, herbicide application, and planting. Even if dominance is not affected, height-growth reductions may result in delayed harvest and this will reduce the return on investments and may interfere with even-flow and sustained yield planning.

Reductions in height growth may also lead to impacts on other forest re-

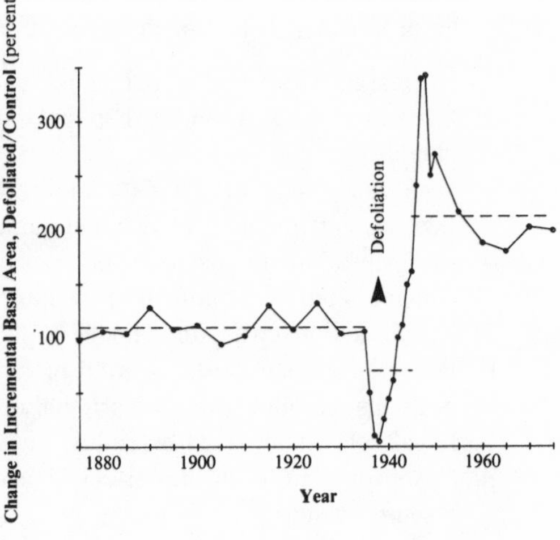

FIGURE 10.3. Average annual basal area growth of white firs heavily defoliated by the Douglas-fir tussock moth in 1936, 1937, and 1938 expressed as a percentage of the growth of nearby undefoliated trees. Note accelerated growth following the defoliation episode. (After Kock and Wickman, 1978.)

sources. Competitive changes in species composition, in particular, can alter land use patterns by big game animals, cattle, and recreationists, and may affect water yields and fire risks. In certain cases, land values may be increased, such as the use of brush fields for big game forage. On the other hand, diameter growth losses should not have serious impacts on forest resource values other than timber production.

10.1.4. Deformity

Infestations by tip and shoot insects, and sometimes by bark beetles and defoliators, may kill the apical shoots or the entire top of the tree and thereby cause serious stem deformities. When this kind of damage occurs to saplings or pole sized trees, lateral shoots usually take over apical dominance and this leads to crooked or forked stems in the mature trees (Fig. 2.2c). Trees with stem deformities yield shorter and smaller diameter sawlogs reducing timber values and yields and sometimes increasing handling costs. Damage to the tops of mature trees does not affect stem form, but it may increase the incidence of heart rot by providing infection courts for decay microorganisms. Except for the visual impact, tip and shoot damage should not have significant effects on other forest resources.

10.2. ESTIMATING THE COSTS OF PEST MANAGEMENT

The other side of the benefit–cost equation (10.1) is the cost of the pest-management operation. From the writings on this subject one might be lead to believe that estimating these costs is a relatively easy task, for it is often assumed that costs are only incurred in the purchase or rental of materials and equipment and the hiring of labor. This assumes, however, that there are no indirect or hidden costs associated with pest management operations, an assumption that is rarely true in forestry. A few of the possible indirect costs are as follows:

1. Decreased water quality due to pesticide contamination or soil erosion
2. Decreased wildlife and fish populations due to poisoning or habitat alteration
3. Decreased range and recreation utilization due to environmental alteration or contamination of fish and game
4. Compensation to landowners, ranchers, and beekeepers for land damage, cattle contamination, or honey-bee poisoning
5. Rehabilitation costs for damage to fisheries and wildlife populations
6. Pest management costs incurred in suppressing outbreaks of secondary pests that resulted from the original operation (see Chapter 9)
7. Legal costs incurred in defending lawsuits
8. Costs for monitoring the effect of management treatments on various resource values.

A meaningful estimate should at least consider these indirect costs. All costs and revenues must also be discounted at the current and expected interest rates over the rotation period.

Finally, before the net benefits of a particular management option can be calculated the probable efficacy of the project needs to be determined. In other words, we need to know the probability that management action will in fact eliminate future damage from the insect outbreak; that is,

$$\text{Values of resources saved} = \text{values of damages expected} \times \text{probability of complete suppression} \quad (10.2)$$

Management efficacy can only be predicted if the effect of the treatment on the long-term dynamics of the pest population is known.

It should be evident by now that pest management decision-making is a complicated and involved process. Informed and economically sound decisions require the prediction of damage expectations over the rotation period, the conversion of damage expectations into monetary losses, perhaps to various forest resources, the determination of the efficacy and costs of various pest management alternatives and, finally, the discounting of revenues and costs over the rotation period. Because of these complicated calculations, computer models are often used to help the manager in the decision-making process.

10.3. DECISION-SUPPORT SYSTEMS

Computer-based decision-support systems are built from a large body of information on the system under management and are designed to permit the manager to test various options in various combinations on the computer rather than on the real ecosystem. This not only saves time and money but permits the manager to learn by his or her mistakes without the drastic consequences that often result from errors in the real world.

Computer models are usually designed to accept inventory data obtained by sampling forest stands and insect or pathogen populations. These data are then used to drive internal predictive models that project stand growth, yield, and damage into the future. Most decision-support models enable the manager to enter a variety of management options, and the model then projects the effects of these activities into the future. For example, the manager may be able to thin the stand or harvest it at different times during the simulation or to initiate fire control or insecticide spray programs. The computer then displays the stand and damage statistics for each management option over the simulation period, and the manager can choose the option that best suits his or her objectives.

One of the more widely used management decision-support systems in the United States is the so-called stand prognosis model developed by scientists at the U.S. Forest Service's Intermountain Forest and Range Experiment Station

(Fig. 10.4). The simulation is initiated by supplying the computer with site, stand, and individual tree inventory data, including habitat type, slope, aspect, and elevation; as well as tree species, diameters, heights, ages, crown lengths, and radial growth increments. The operator then specifies certain management options such as thinnings and harvest times. The computer calculates the stand growth and yield statistics for a constant time interval, say every 10 years, and displays stand attributes at the end of each interval; e.g., species composition, diameter distribution, age distribution, timber volumes, and volumes added and lost due to growth and mortality. These computations are then repeated for the desired number of intervals.

The stand prognosis model can interact with several insect population models; e.g., submodels are currently available for the mountain pine beetle (see Fig. 6.10), the Douglas-fir tussock moth, and the western spruce budworm. Insect pest management treatments such as insecticide or virus sprays can be implemented against the defoliating insects (Fig. 10.5a). Alternatively, variables generated by the stand prognosis model can be displayed on a risk graph such as that developed in Chapter 7. This graphical display shows the danger periods during stand development when the likelihood of insect outbreaks becomes extreme (Fig. 10.5b).

Data generated by the stand prognosis model can also be passed to various other submodels that forecast water yields and sediment transport, elk and moose forage yields, and fish production. Through these submodels, the impacts of timber harvesting and insect outbreaks on water, wildlife, and fish production can be estimated. In addition, the output from the stand prognosis and its adjunct submodels can be transmitted to an economic analysis program that estimates the *net present value* of various forest resources and uses. For example, the net present values for timber volumes projected in Fig. 10.5a are given in Fig. 10.6. We see that, although the greatest volume yields resulted from outbreak suppression with insecticides (Fig. 10.5a), the greatest values were recovered under the two thinning options (Fig. 10.6). This was because insecticide treatment preserved the host trees and gave rise to a small diameter, highly susceptible stand. By contrast, the thinnings produced a resistant stand of valuable large-diameter trees and also captured income from commercial thinnings.

The critical components of any prognosis model are the different population growth submodels. These models are usually constructed from a large body of empirical data, as well as theoretical and intuitive knowledge, and are responsible for predicting changes in the populations of trees, insects, wildlife, and so forth. Because biological systems are rarely understood completely, and because mathematical descriptions of biological processes may be incorrectly formulated or joined together, we should expect imperfections in the model projections. In order to check the model forecasts, the natural system must be carefully monitored (Chapter 6). On the basis of comparisons between model predictions and the real world, the model is re-evaluated and reformulated so that it more accu-

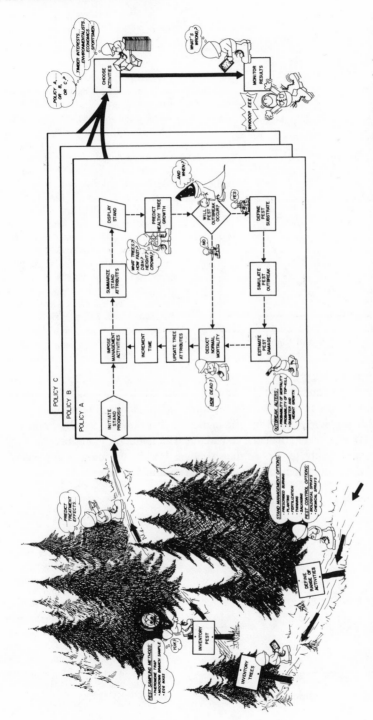

FIGURE 10.4. Elementary schematic of the stand prognosis model. The model is initiated with data from stand and pest inventories and a set of management activities. It then predicts tree and insect population growth and displays a summary of new stand attributes over a series of equal time periods. The output is used to choose certain management alternatives. The results of these activities are then monitored in the field to check the model forecasts. (Courtesy of USDA Forest Service, Intermountain Forest and Range Experiment Station.)

CHAPTER 10

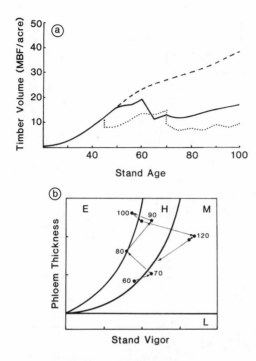

FIGURE 10.5. Results of simulations by the stand prognosis model. (a) Projected timber volume per acre in a mixed Douglas-fir–grand fir stand subjected to five outbreaks of the Douglas-fir tussock moth and various management options. (—) no treatment; (---) insecticide application to each outbreak, (...) one precommercial and one commercial thinning. (From May *et al.*, 1984.) (b) The 10-year incremental trajectory of a simulated lodgepole pine stand showing changing risk of mountain pine beetle outbreaks. E, extreme risk; H, high; M, moderate; and L, low risk. Each point from 70 to 140 represents the phloem thickness distribution and stand resistance at the end of the 10-year interval. See Chapter 7 for a detailed description of the risk model. (From Berryman, 1978.)

rately reflects the actual situation. Through this feedback process of evaluation and adjustment, the performance of the model is gradually improved.

Another problem with prognosis models is that they are usually limited in scale because of time and financial constraints. Computer programs become time consuming and costly to run as they become bigger and more complicated. Because of this, some components of the real system may have to be omitted or

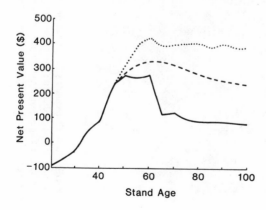

FIGURE 10.6. Projected net present value per acre of a mixed Douglas-fir–grand fir stand subjected to five Douglas-fir tussock moth outbreaks and various management options (—) no treatment; (---) insecticide treatment of each outbreak; (...) one precommercial and one commercial thinning. Net present value equals the present value of stand, i.e., timber only minus management costs. (From May *et al.* 1984.)

abstracted. If missing components have important effects on the behavior of the populations we are interested in, the model may give erroneous projections. For example, birds may be important in regulating some forest insect populations, and if bird population models are not part of the decision-support system, then the effects of insecticide treatments on these predators cannot be predicted. This may give rise to erroneous conclusions concerning the benefits of insecticide applications.

In the long run, decision-support systems will probably prove an invaluable aid to the forest manager, permitting various management options in various combinations to be tried on the modeled system without the costs and risks associated with experiments in real life. Models, however, are only abstractions of the real world, and their predictions should be treated with some caution. Models, no matter how sophisticated, will never replace the sound judgment that results from an intimate understanding of ecological relationships among trees, insects, and the other members of the complex ecosystem in which we work.

10.4. INTEGRATED PEST MANAGEMENT

One of the major advantages of a well-designed decision-support system is the ability to test a variety of pest management alternatives, either singly or in combination, in the environment of the model ecosystem. Because this enables the manager to observe the effect of integrating several management options, the decision-support system has become the heart of what is commonly called an *integrated pest management program.*

Arising in Europe under the label *harmonious control,* the objective of an integrated pest management program is to minimize the damage caused by insects with a carefully designed blend of control procedures that have as little impact on the other components of the ecosystem as possible. In order to achieve these objectives, a great deal of information is needed on the interrelationships between the components of the managed ecosystem. A model attempts to bring this information together in a holistic manner so that the manager can view the ecological and economic consequences of control actions.

The adoption of integrated control was particularly spurred by concerns about the widespread and excessive use of chemical pesticides and their effects on ecological systems and human health (Chapter 9). Although most integrated pest management specialists acknowledge that pesticides are important components of the manager's arsenal, they usually view pesticides as a last resort, to be used in a manner that minimizes disruption to natural controls and other ecological processes. Nowadays, integrated control is often supplemented by mathematical *optimization* methods that calculate the *blend* of control tactics that optimizes the management objective function. The optimal control strategy may consist of a series of silvicultural and natural enemy manipulations together with an occasional tactical pesticide application, i.e., when the other natural controls fail. This strategy is obtained by evaluating a number of control options, singly and in combination, to arrive at an optimal combination which minimizes the impact of insects on one or several forest resources. Because of the large number of options and objectives that may have to be considered, as well as the complex and tedious mathematics involved, computers are an essential part of the optimization process. This is particularly true in forestry, where the natural ecosystem is composed of a large number of interdependent plants and animals. Even though these models are rarely perfect, they at least provide us with a rational and objective procedure for making pest control decisions. They should not, however, replace that adaptive and flexible component of the pest management system, the human mind. The manager must never be hesitant to innovate, to take risks when they will provide the information to improve later management decisions. No matter how good our models are, the natural ecosystem will always present us with surprises and will sometimes react in unexpected ways to our manipulations.

Adaptive managers will be able to learn from these surprises, will change their minds, will update their mental and mathematical models. They will end up with a better understanding of the real world, and their models will record this understanding so that their successors will benefit from their mistakes. The idea of an adaptive component to the management system, the human mind, has given rise to a new approach in decision-making called *adaptive management,* in which trial-and-error are treated as an integral component of the management decision-making process, and the unexpected becomes expected. The formalization of adaptive management procedures is, perhaps, the wave of the future.

REFERENCES AND SELECTED READINGS

Berryman, A. A., and Pienaar, L. V., 1974, Simulation: A powerful method of investigating the dynamics and management of insect populations, *Environ. Entomol.* 3:199–207.

Berryman, A. A., 1978, A synoptic model of the lodgepole pine/mountain pine beetle interaction and its potential application in forest management, in: Theory and Practice of Mountain Pine Beetle Management in Lodgepole Pine Forests (A. A. Berryman, G. D. Amman, R. W. Stark and D. L. Kibbee, eds.), pp. 95–108, University of Idaho, Forestry, Wildlife and Range Experiment Station, Moscow.

Bethlahmy, N., 1975, A Colorado episode: Beetle epidemic, ghost forests, more streamflow, *NW Sci.* 49:95–105. (watershed yields)

Downing, K. B., and Williams, W. R., 1978, Douglas-fir tussock moth: Did it affect private recreational businesses in northeastern Oregon?, *J. For.* 76:29–30. (economic impact)

Graham, K., 1963, *Concepts of Forest Entomology,* Reinhold Biological Sciences, New York. (see Chapter 5, Economic Evaluation)

Holling, C. S. (ed.), 1978, *Adaptive Environment Assessment and Management,* International Series on Applied Systems Analysis, Wiley, Toronto, 377 pp.

Klock, G. O., and Wickman, B. E., 1978, Ecosystem effects, in: The Douglas-Fir Tussock Moth: A Synthesis, (M. H. Brookes, R. W. Stark, and R. W. Campbell, eds.), pp. 90–95, Technical Bulletin 1585, U.S. Forest Service. (impact of Douglas-fir tussock moth)

Kulman, H. M., 1971, Effects of insect defoliation on growth and mortality of trees, *Annu. Rev. Entomol.* **16:**289–324.

Leuschner, W. A., 1978, Elements of a typical IPM system: The socio-economic and decision-making model, Proceedings of the Society of American Foresters Convention, St. Louis, Missouri, pp. 263–267.

Leuschner, W. A., and Berck, P., 1985, Decision analysis, in: *Integrated Pest Management in Pine–Bark Beetle Ecosystems* (W. E. Waters, R. W. Stark, and D. L. Wood, eds.), pp. 177–189, Wiley (Interscience), New York.

Leuschner, W. A., Matney, T. G., and Burkhart, H. E., 1977, Simulating southern pine beetle activity for pest management decisions, *Can. J. For. Res.* **7:**138–144.

Marshall, K. B., 1975, The spruce budworm and the dollar in New Brunswick, *For. Chron.* **51**(4):9–12. (economic impact)

May, D. M., Stozek, K. J., and Dewey, J. E., 1984, A demonstration to risk-rate stands to Douglas-fir tussock moth defoliation on the Palouse Ranger District, Clearwater National Forest, Idaho. 2. Predicting defoliation levels and evaluating alternative control treatments, U.S. Forest Service, Northern Region, State and Private Forestry, Report Number 84-6.

Monserud, R. A., 1978, Combining the stand prognosis and Douglas-fir tussock moth outbreak models, in: Proceedings of the Society of American Foresters Convention, St. Louis, Missouri, pp. 268–272.

Monserud, R. A., and Crookston, N. L., 1982, A user's guide to the combined stand prognosis and Douglas-fir tussock moth outbreak model, U.S. Forest Service General Technical Report INT-127, 49 pp.

Stage, A. R., 1973, Prognosis model for stand development, U.S. Forest Service Research Paper INT-137, 32 pp.

Stage, A. R., Babcock, R. K., and Wykoff, W. R., 1980, Stand-orientated inventory and growth projection methods improve harvest scheduling on Bitterroot National Forest, *J. For.* **78:**265–278.

Stevens, R. E., and Jennings, D. T., 1975, Western pine-shoot borer: A threat to intensive management of ponderosa pine in the Rocky Mountain area and southwest, U.S. Forest Service, General Technical Report RM-45, 8 pp.

Stoszek, K. J., 1973, Damage to ponderosa pine plantations by the western pine-shoot borer, *J. For.* **71:**701–705. (measuring impact of height reduction due to shoot damage)

Waters, W. E., and Stark, R. W., 1980, Forest pest management: Concept and reality, *Annu. Rev. Entomol.* **25:**479–509.

Waters, W. E., Stark, R. W., and Wood, D. L., 1985, *Integrated Pest Management in Pine–Bark Beetle Ecosystems,* Wiley (Interscience), New York.

Wykoff, W. R., Crookston, N. L., and Stage, A. R., 1982, User's guide to the stand prognosis model, U.S. Forest Service General Technical Report INT-133, 112 pp.

PART III: MANAGEMENT EXERCISES

III.1. A survey of three ponderosa pine stands was designed to determine the status of mountain pine beetle populations. On each infested tree along a linear transect, two 6 × 6-in. bark samples were removed and the total number of beetle larvae, pupae, and adults was counted. The total counts from each tree are shown for three stands:

Stand 1 = 2 beetles (tree 1), 1 (tree 2), 0, 2, 1, 0, 1, 0, 1, 0.
Stand 2 = 10, 18, 22, 8, 41, 43, 80, 14, 39, 21.
Stand 3 = 21, 34, 9, 15, 10, 36, 6, 18, 12, 21.

Use the sequential sampling plan in Fig. 6.8 to decide on the status of these beetle populations. What would you need to do in stand 3?

III.2. In New Brunswick the black-headed budworm, a defoliator of fir and spruce, was measured in 1947 and 1948 (Morris, 1959) by clipping foliage samples from host trees to provide the following estimates: 1947 = 112 larvae/100 ft^2 foliage; 1948 = 533 larvae/100 ft^2 foliage.

a. Calculate the trend index [Eq. (6.2)] of the budworm population over this period. What does this index tell you about the budworm population in 1949? How much reliance would you place on this prediction? Why?

b. It was also found that 230 budworms were parasitized in 1948. Use Fig. 6.7 and Eq. (6.5) to determine the trend of the population next year. How much reliance would you place on this prediction? Why?

c. The actual density of budworm larvae in 1949 was 225 larvae/100 ft^2 of foliage. Calculate the real trend index.

III.3. An increment core from a pine tree provided the following data: tree radius = 25 cm, sapwood depth = 10 cm, bark thickness = 0.7 cm, and annual increments from the present to 10 years ago = 0.2, 0.3, 0.2, 0.2, 0.3, 0.3, 0.4, 0.6, 0.5, and 0.7 cm. Calculate the 5-year periodic growth ratio and relative sapwood increment (see Fig. 7.2).

III.4. Using Table 7.4, determine the risk of a spruce beetle outbreak in a pure spruce stand with site index 100, basal area 180 ft^2/acre, and mean diameter breast height of 20 in.

III.5. Using Table 7.5, assess the risk of a ponderosa pine being killed by the western pine beetle when it has normal needle complement and color but needles are short in the top of the crown and there are no dead twigs or branches or other apparent injuries.

III.6. Using the equation in Table 7.6 (a), assess the probability of a white fir tree dying from bark beetle attack when the branches are all drooping downward, 30% of the branches are dead or fading and no live bark is visible on the main trunk.

III.7. Using the equations in Table 7.6 (b), assess the risk of southern pine beetle infestation in a pure loblolly pine stand with a basal area of 120 ft^2/acre, average age 40 years, and average 10-year radial increment 0.2 in.

III.8. Using Fig. 7.4, assess the risk of a mountain pine beetle outbreak occurring in a 72-year-old pure lodgepole pine stand when the average 10-year growth increment is 0.4 in., the average 5-year growth increment is 0.25 in, the crown competition factor is 92, and 60% of the stand basal area has phloem greater than $1/10$ of an inch thick.

III.9. Using the equation given in example 1 of Table 7.7, calculate the expected percentage defoliation of a grand fir stand with the following characteristics: upper slope, 1 m of volcanic ash in the soil, average age of host trees 100 years, stand basal area 100 m^2/ha, site index 40 m at 50 years, and only grand fir present in the canopy.

III.10. The Douglas-fir tussock moth defoliated vast areas of the Blue Mountains of Oregon and Washington in 1973. In order to justify the control of the outbreak with chemical pesticides, an economic analysis was performed by a Forest Pest Management Task Force (USDA–USDA Environmental Statement, Cooperative Douglas-fir Tussock Moth Pest Management Plan, Idaho–Oregon–Washington, December 1973). On the basis of the assessment of

defoliation damage in 1973, the task force predicted that the damage expected in 1974 would
be as follows (figures rounded by the author):

Acres infested	331,000
Mortality MMBF*	656
Growth impact MMBF	240
Salvage potential MMBF	470
Stumpage value	$56/M* for green timber;
	$40/M for infested timber

*Note that M or MBF means thousand board-feet, and MMBF means million board feet.

The task force also estimated additional expenses and/or benefits as follows:

Reforestation (e.g., planting, site preparation: in killed or salvaged areas)	$ 2,630,000
Fire protection (increased surveillance, constructing fuel breaks, and fuel reduction costs)	13,130,000
Wildlife habitat (no value was assigned)	0
Recreational usage (reduction of tourism, hunting, and fishing in visitor-days)	1,460,000
Range utilization (increased forage production, a benefit)	101,000

The costs of spraying the total outbreak area with the insecticide DDT was estimated at
$4.00/acre based on insecticide and fuel oil prices, costs of chemicals and aircraft, salaries
and travel expenses, and equipment rental.

As DDT is very effective against the tussock moth, we can assume that the spray program
would completely protect the forest from defoliation damage in 1974. On this assumption,
calculate the net benefit of the proposed control project, then criticize the figures and
assumptions used in this analysis, particularly those relating to growth impact, wildlife
habitat, recreation, and the costs of the control operation.

III.11. Using the insect chosen for your special project:
   a. Describe, from the literature, methods used for monitoring, forecasting and assessing the
      risk of outbreaks. Also discuss the history of prevention and control and whether simula-
      tion models are available.
   b. From your readings on this insect, discuss the opportunities for improving the current
      monitoring, forecasting, risk assessment, prevention, control, and decision-making pro-
      cedures. List what you consider the highest priorities for future research.
   c. Design what you believe will be an effective integrated pest management strategy for this
      pest, using the published literature and your own imagination.

PART IV

PRACTICE

"Theory without practice is fantasy, practice without theory is chaos."

Abridged from a forgotten author.

Photographed by M. M. Furniss, USDA Forest Service, retired.

CHAPTER 11

HE PRACTICE OF FOREST INSECT MANAGEMENT

Throughout this book I have attempted to develop a general framework for understanding and managing forest insect pest populations—general in the sense that the principles and methods should be applicable to any insect pest problem anywhere in the world. I had originally planned to end the book at Chapter 10 but, on reaching that point, I became concerned that the reader would not have grasped how these principles and methods could be applied in practice. Even though the exercises at the end of each section, particularly the special project, were designed to provide experience in solving forest insect problems, nowhere is the practical side of forest management consolidated into a unifying whole.

Having decided that a chapter on the practice of forest insect management was desirable, I was immediately confronted by the problem of how to organize and write such a chapter. Practice by definition means doing something, not passively reading how someone else would do it. In addition, there are often as many ways of solving a particular problem as there are people—practice involves individual preferences and, although my approach may be different from yours, it is not necessarily better!

Earlier in this book I likened the forest pest manager to a medical doctor and, indeed, their professions are remarkably similar. Both are concerned with the health of living systems, and both approach the problem through diagnosis, prognosis, and prescription.

Diagnosis is the art of recognizing the presence and extent of a disease or insect infestation and determining the causal agent. The forest pest manager is constantly on the lookout for *symptoms* of insect activity, such as gray (defoliated) trees, yellow or red-tops, crooked stems, and forked tops. A careful examination of these symptoms not only enables the diagnostician to detect the presence and extent of insect damage but may also provide clues to the species of insect involved. In cases in which the pest does not have a local reputation, however, the advice of a specialist must be sought to obtain a specific identification. The correct diagnosis of forest insect problems depends on the accuracy of the monitoring techniques (Chapter 6), the experience of the diagnostician, and the availability of specialists.

Prognosis is the art of anticipating problems before they occur and of predicting the probable course and seriousness of specific problems after they have been diagnosed. As a doctor tries to anticipate the likelihood of a heart attack based on the patient's job, diet, smoking habits, and blood pressure, so a silviculturist attempts to assess the probability of a stand being damaged by insects on the basis of an analysis of either stand or site factors, or both (Chapter 7). Although formal risk assessment models are helpful in this effort, informal assessments based on the general health and vigor of the stand are useful as well. In general, unhealthy stands growing on poor sites or under crowded conditions are most likely to be damaged by insect infestations.

Predicting the course of ongoing insect infestations requires information on the population dynamics of the insect in question (Chapter 4). If the pest exhibits gradient outbreaks, for example, we would expect the infestations to be short-lived and/or restricted to specific sites. Eruptive outbreaks, on the other hand, are likely to spread over large areas, to last for a long time, and to cause more serious damage to the forest resources. Forecasting models may be helpful in predicting the course and seriousness of insect outbreaks (Chapter 6) but do not substitute for a keen understanding of the population dynamics of the pest organism.

Prescriptions in both forestry and medicine can be either preventive or therapeutic. Preventive forest pest management, like preventive medicine, attempts to treat the patient before a disease condition occurs. In order to prescribe preventive treatments, the pest manager must be able to evaluate the health of forest stands and the risk of insect outbreaks (Chapter 7). In addition, because preventive treatments usually involve silvicultural or natural enemy manipulations, the pest manager needs a general understanding of the influence of these factors on the population dynamics of the pest (Chapters 5 and 8).

Therapeutic treatments are usually prescribed if the prognosis for a current outbreak indicates a serious and deteriorating condition. The goal of therapy is to contain or suppress the pest population, but successful treatments also require information on the population dynamics of the pest species, as well as the effects of the various treatments on the population (Chapter 9).

From the previous discussion it is apparent that the forest pest management practitioner must have a minimum set of information in order to deal with specific forest insect problems:

1. *The identity of the insect causing the problem:* This information opens up the scientific literature to scrutiny for the following information.

2. *The biology, behavior, life cycle, and ecology of the pest:* In particular, it is very important to try and identify the type of population dynamics (gradient or eruptive) and to understand how these dynamics are influence by interactions with the food plant and natural enemies.

3. *The appropriate tools for managing the pest:* Formal and/or informal methods for evaluating the risk of pest outbreaks, techniques for monitoring and forecasting infestations, information on the probable effects of silvicultural and natural enemy manipulations on the pest population, and the types of therapeutic treatments effective for containing and suppressing outbreaks. Once all this information has been gathered, it should be possible to design a comprehensive management plan for the particular pest.

This chapter will attempt to develop management practices for specific forest insects. The first example attacks the problem of bark beetles infesting spruce forests in Norway; the relevant information is assembled, the ecological interactions are defined, the population dynamics are classified, and diagnostic, prognostic, prescriptive, and management methodologies are developed. The next two examples supply relevant data, but their application is left to the student. Finally, for students who wish more practice, or for teachers who wish to assign problems to students, a list of forest insects is supplied together with key publications. In most cases there is sufficient information on the insects involved to prepare a reasonable practical analysis.

11.1. THE SPRUCE BARK BEETLE IN NORWAY

The spruce bark beetle, *Ips typographus* L. (Coleoptera: Scolytidae), is the most destructive bark beetle in the coniferous forests of Eurasia. The adult beetles (Fig. 11.1) overwinter in the litter on the forest floor or in the bark of spruce trees and fly in early spring as soon as the weather warms sufficiently (mid to late May). Individual males attack Norway spruce trees or logs, *Picea abies* (L.) Karst. and release powerful aggregating pheromones that attract both males and females to the attacked tree. The male is polygamous and may be joined by up to five females, each of which bores an individual egg gallery in the phloem and deposits eggs in niches cut in the lateral walls (Fig. 11.1). If the brood tree is crowded, parent beetles may leave and produce one or two sister broods in other trees (June–July). Upon eclosion, the larvae feed on the phloem tissue and eventually pupate at the end of their feeding galleries. The adult beetles then feed for a short time beneath the bark before boring out and moving to the litter at the base of the tree, where they overwinter. About 10% of the adults spend winter in the bark of host trees. In Scandinavia, generally a single generation is produced each year, whereas two or even three generations may be produced in Southern Europe.

FIGURE 11.1. Life cycle of *Ips typographus* in southern Norway. (a) Adult beetles overwinter in the litter and under bark. (b) They emerge and attack trees, logs or windthrows in early spring. (c) Parent beetles may reemerge in summer and attack new trees or logs. (d) Brood beetles emerge in autumn and enter overwintering sites.

11.1.1. Interactions with the Tree

Like most other bark beetles, *Ips typographus* adults introduce several fungi into the tree during their boring beneath the bark. One of these species, *Ceratocystis polonica* (Siem.) C. Moreau, has been shown to be highly pathogenic. In fact, healthy Norway spruce trees can be killed by this fungus alone when the tree is artificially inoculated at multiple points around its circumference (150–200 inocula frequently kill healthy spruce trees of 20-cm diameter).

Spruce trees are able to defend themselves against beetle attacks. Although they do not have the extensive resin duct system of pines, some resin flow usually occurs from severed ducts as soon as the beetle enters the phloem. The tree also reacts to the invasion of pathogenic fungi with a typical hypersensitive wound response in the phloem and sapwood (see Chapter 3, Fig. 3.1). If the tree is healthy, these defensive responses will normally repel the beetle and contain the spread of the pathogen, provided the beetle attack is not too dense.

The spruce beetle aggregation pheromones, methylbutenol, *cis*-verbenol, and ipsdienol, are oxidation products of host terpenes, particularly α-pinene and myrcene. These products form an extremely potent mixture that enables beetles to overwhelm the resistance of their hosts if sufficient beetles are present in the vicinity to respond to the pheromone.

Pheromone mixtures incorporated into plastic strips or placed within plastic bags can also be used in pipe traps (Chapter 9, Fig. 9.5) to capture large numbers of flying beetles. When used for suppression, the traps should always be placed in open areas and at some distance from the forest edge, preferably in clearcuts of the previous year. This is done to prevent nearby trees from being attacked. Pheromone traps also capture clerid predators (*Thanasimus* spp.) at a rate of 1 clerid for every 1000 bark beetles.

11.1.2. Interactions with Natural Enemies

Although *Ips typographus* is attacked by numerous insect predators and parasitoids, as well as woodpeckers, there is no evidence that they are capable of regulating spruce bark beetle populations when susceptible hosts are available. There seems to be little opportunity at the present time for augmenting natural enemy populations. Even woodpeckers offer little hope because few *Ips* adults are present in the tree during winter, when woodpeckers usually feed heavily on bark beetles. Nevertheless, there may be opportunities for preserving natural enemies during suppression projects. Even though they do not seem able to regulate spruce bark beetle populations, natural enemies may slow down the rate of population increase and accelerate the rate of decline.

11.1.3. Outbreak History

Outbreaks of spruce bark beetles have occurred at irregular intervals throughout the coniferous forests of Eurasia but are usually associated with management neglect or environmental disturbances. For example, an epidemic in Norway during the 1850s was preceded by storm damage and a severe drought, while that in Germany during the 1940s occurred in forests neglected during the war and damaged in the fighting.

The recent spruce beetle outbreak in southern Norway began in 1971 in south Hedmark County (Area 1 in Fig. 11.2). This area had suffered heavy spruce blowdown during the gale of November 1969. The windthrown trees provided breeding material for *Ips* beetles in 1970, and the brood emerging in 1971 attacked standing trees. Although moderate to severe blowdown also occurred in a strip running along the coast of Area 3 (Fig. 11.2), spruce beetle populations did not increase appreciably in this area. A much slower buildup was evident in area 2, which also suffered moderate to severe blowdown.

Storms in 1975 and 1976 felled numbers of spruce trees, particularly in area 3 to the southwest of Oslo, but the blowdown was relatively modest compared to the gale of 1969. The same years, however, parts of Europe experienced the worst drought in recorded history. In southeastern Norway summer rainfall during 1974, 1975, and 1976 was only 30–60% of normal, the worst affected regions being areas 2 and 3. During the years after the drought, spruce mortality due to beetle attack increased in both areas but was most extreme in area 3.

The analysis of data from damage assessment surveys indicated that heaviest spruce mortality was associated with low elevation (<600 m) sites that had thin soils and heavily fissured (well-drained) bedrock. Damage was also more severe on valley sides, particularly steep north and east facing slopes, and on high or medium site classes. Regions with large continuous areas of spruce were also more heavily attacked.

11.1.4. Conventional Management Practices

In the intensively managed spruce forests of central Europe, the normal practice for dealing with *Ips typographus* is through silvicultural treatments to maintain stand vigor, as well as sanitation to eliminate infested trees and logging debris. In the past, infestation of newly harvested timber was prevented by peeling the bark from all logs in the forest. In many countries, including Norway, this practice was abandoned during the 1960s for economic reasons (labor intensive and high cost). In some European countries, *Ips* populations are also managed by the trap-tree method (Chapter 9) with 5 trap trees usually being felled for each infested spruce. Trap trees are then removed and treated after they become infested with spruce beetles.

The conditions in many parts of southeastern Norway hinder the use of intensive management techniques. The forests are mostly privately owned and often split up into small woodlots. The terrain, particularly in area 3, is very steep and many regions are quite inaccessible. Small owners often use their forest holdings as savings to be used in the case of financial emergency. Thus, many stands are neglected and tend to be overstocked and overaged.

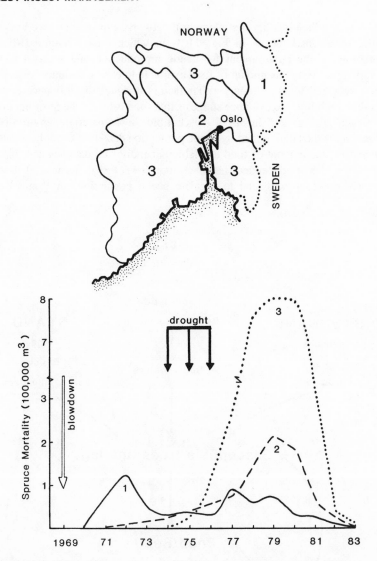

FIGURE 11.2. Historical aspects of the *Ips typographus* outbreak in southeastern Norway, 1971–1983. (From Worrell, 1983; and Professor A. Bakke, Norwegian Forest Research Institute, personal communication.)

11.1.5. Understanding the Dynamics of Spruce Beetle Populations

In this, the first step of our practical analysis, we attempted to identify the structural and dynamic properties of the spruce bark beetle population system in order to understand why it behaves as it does. We need to identify the main environmental variables within the conceptual structure developed in Chapter 4

(Fig. 4.1) and then identify the interaction network. I have attempted to do this for the spruce bark beetle in Fig. 11.3, in which I have assumed that the favorability of the environment for beetle reproduction and survival is most strongly affected by the amount of food available to each individual (susceptible spruce trees and logs). Although other factors undoubtedly influence environmental favorability, the evidence suggests that food is by far the most important.

Given a constant input of susceptible host material, environmental favorability is then determined by beetle population density; i.e., as beetle density increases, the amount of food available per individual declines and this produces a (−) density-dependent feedback loop: $(+f)\cdot(-g)$. Because of its rapid response, this loop will tend to regulate beetle populations in "tight" equi-

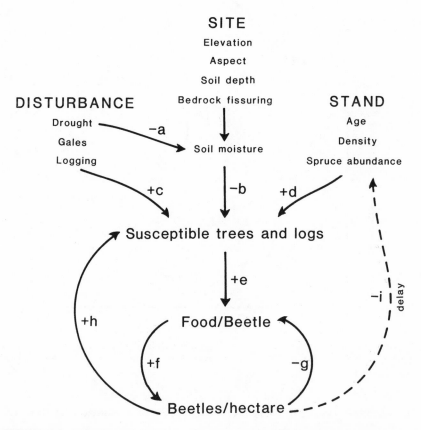

FIGURE 11.3. Feedback analysis of the spruce bark beetle population system. The basic environment is divided into site, stand, and disturbance variables; the favorability of the environment for individual beetles is directly related to the amount of food available and the density of the beetle population. See text for explanation.

librium when their food supplies remain constant. The equilibrium density is completely determined by the quantity of susceptible host material.

I have identified three sets of basic environmental variables that determine the quantity of susceptible spruces available to the beetles. Obviously the local abundance of spruce trees and the density of the stand influences this quantity, as does stand age (older trees normally being weaker). Two of these variables, spruce abundance and density, are also affected by beetle population density, particularly when the death rate of trees due to the beetle attack exceeds the growth rate of the stand. Hence we have a second $(-)$ density-dependent feedback loop: $(-i)\cdot(+d)\cdot(+e)\cdot(+f)$. Because this loop operates through the basic environment, however, it will have a time delay; i.e., the food supply of future beetle generations is affected because suitable host material regenerates very slowly.

The number of susceptible spruce trees and logs is also affected by site factors, particularly soil moisture. It is also affected by disturbances such as logging, windstorms, and droughts.

Finally, the evidence suggests that dense spruce beetle populations can increase the number of susceptible hosts in a similar manner to that displayed by some other bark beetles (see Chapter 4, Fig. 4.5). This interaction gives rise to a $(+)$ density-dependent feedback loop, i.e., $(+h)\cdot(+e)\cdot(+f)$, which, we know, can result in an unstable outbreak threshold. The spruce beetle, should therefore be capable of eruptive population growth.

The next step in the analysis is to identify the potential equilibrium points in the system by deducing the form of the individual birth and death rate curves. We start by setting the input of susceptible spruce trees and logs at a constant rate, say two diseased or felled trees per hectare per year. We then imagine how the individual death rate will change as population density rises (Fig. 11.4). As less food will be available per individual as density increases, the death rate should rise at first; i.e., the $(-)$ feedback loop $(+f)\cdot(-g)$ is in operation. The birth rate

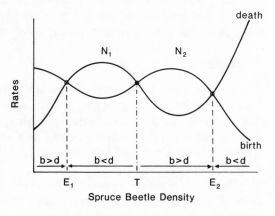

FIGURE 11.4. Deduction of the individual birth and death rate curves for the spruce bark beetle showing a low-density stable equilibrium E_1, an unstable outbreak threshold T, and a high-density cyclic equilibrium E_2. See text for explanation.

may also decline as individual females abandon the overcrowded trees before depositing their full egg complement. As the population becomes even more dense, however, they will be able to overcome some of the more healthy spruces, leading to an increased quantity of susceptible host material. As a result of this (+) density-dependent feedback, i.e., loop $(+h)\cdot(+e)\cdot(+f)$, the individual death rate will fall, while the birth rate may rise after the population reaches a certain density (i.e., when $N > N_1$ in Fig. 11.4).

Eventually the expanding population will become so dense that the stand will be severely depleted, i.e., loop $(+d)\cdot(+e)\cdot(+f)\cdot(-i)$ comes into effect, and, at N_2, the death rate will rise again, while births will probably decline.

We now see that the spruce bark beetle system has three potential equilibrium points (Fig. 11.4): E_1, a low-density stable equilibrium; T, and unstable outbreak threshold; and E_2, a high-density cyclic equilibrium, because of the delay in loop $(+d)\cdot(+e)\cdot(+f)\cdot(-i)$. Furthermore, we can see that outbreaks can be triggered by the following factors:

1. Increasing the number of severely weakened spruce trees (logging or windthrows), which raises E_1 toward T (outbreaks are triggered when $E_1 > T$)
2. Decreasing the overall level of resistance in the stand (low soil moisture and old, dense stands) which lowers T toward E_1
3. An immigration of beetles from surrounding areas, raising the endemic population from E_1 to T (i.e., $E_1 + I > T$, where I is the number of immigrants).

This analysis sets the stage for attacking the remaining problems.

11.1.6. Diagnosis

Spruce beetle outbreaks are easily diagnosed because infested trees turn orange and drop their needles a few weeks to several months after attack, and dead trees are often clumped into groups of several to many individuals. Likewise, because successfully attacked trees are killed, timber volume losses are easily estimated from aerial photography or ground surveys.

11.1.7. Prognosis

Because spruce beetle outbreaks are of the eruptive type, we should expect them to intensify and spread as long as susceptible trees remain in the forest. We should also suspect the rate of timber mortality to decline in time as susceptible trees are killed and the survivors recover their resistance (the thinning effect). Weather factors such as cool wet summers and cold dry winters may also hasten the collapse of outbreaks.

The likelihood of spruce beetle outbreaks starting in particular stands can be assessed, informally, by careful examination of certain key variables. As stand health or vigor (resistance) is probably the most important variable, we should be able to assess stand resistance by taking increment cores from a sample of trees and then calculating the periodic growth ratio (PGR) or relative sapwood increment (RSI) (Chapter 7, Fig. 7.2). These measurements integrate the effect of all stress factors acting on the stand, rather like taking the temperature of the stand; a PGR > 1.5 or an RSI > 0.07 usually means that the stand is in pretty good health. Alternatively one could try to obtain an indirect measurement of stand vigor by assessing site factors (soil depth, bedrock fissuring, site class, slope, and aspect), stand density, and spruce abundance (see Section 11.1.3).

Although stand vigor is an important predictive variable, we know that eruptive population systems are also sensitive to insect numbers. The manager should therefore attempt to estimate the resident beetle population inhabiting weakened trees or logging debris, as well as the proximity of beetle infestations. Consideration of all these factors can permit a reasonable prognosis of the likelihood of a spruce bark beetle outbreak starting in a particular area.

11.1.8. Prescription

Prescriptions should always be based on a knowledge of the population dynamics of the insect and a prognosis of the probable course of an infestation. For example, a manager might prescribe low thinning as a preventative treatment if the stand is overstocked (provided thinning is done several years before infestation), clear or shelterwood cutting if the stand is unlikely to respond to thinning (too old or stagnated), selective logging of spruce and encouragement of other species (e.g., Scots pine) if the site is too dry, or sanitation-salvage cuts to remove diseased or damaged trees. If preventative silvicultural practices are impossible, because of costs, terrain or the immediacy of a current outbreak, we may wish to suppress beetle populations with pheromone traps. Suppression by this method seems to be a viable tactic with spruce bark beetles because the pheromone is highly attractive and the traps effective. Suppression is also possible by harvesting trees containing beetle broods provided they are processed before the beetles emerge. Remember that suppression will be most successful in the early stages of an eruptive outbreak if the population can be reduced below the unstable outbreak threshold over the whole infested area. Pheromones could also be used to contain a spruce beetle infestation by baiting living trees in a circle around a spot infestation or along the front of an expanding infestation. We should also note that the option to do nothing in the face of bark beetle population eruptions is extremely dangerous because the outbreak can spread over extremely large areas, including other ownerships. If no action is taken, the manager becomes vulnerable to lawsuit.

11.1.9. Management Planning

A management plan should be a logical step-by-step procedure for diagnosing, prognosticating, and prescribing treatments, as well as for carrying out treatment operations (Fig. 11.5). In this particular plan, the manager has the

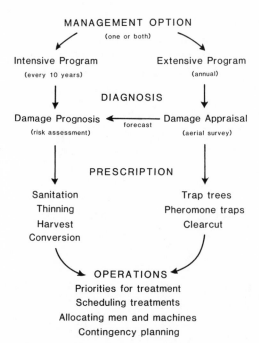

MANAGEMENT OPTION
(one or both)

Intensive Program   Extensive Program
(every 10 years)   (annual)

DIAGNOSIS

Damage Prognosis ←——————— Damage Appraisal
(risk assessment)   forecast   (aerial survey)

PRESCRIPTION

Sanitation   Trap trees
Thinning   Pheromone traps
Harvest   Clearcut
Conversion

OPERATIONS
Priorities for treatment
Scheduling treatments
Allocating men and machines
Contingency planning

FIGURE 11.5. Step-by-step management plan for the spruce bark beetle in Norway spruce forests.

option of either an intensive program involving individual stand analysis or an extensive program involving aerial survey. The forest manager may wish to execute either the intensive program or both programs in accessible or high-value areas, reserving the extensive program for inaccessible or low-value areas or for areas not used for timber production.

The intensive program involves individual stand inspection and formal or informal prognosis of the risk of spruce beetle outbreaks (see Section 11.1.7). Individual stands are then allocated to the treatment or to no treatment categories, on the basis of treatment priorities. For example, medium risk stands may be allocated to the no-treatment block because of insufficient resources or time to treat all blocks. High-risk stands would almost always be treated because of the danger of eruptive outbreaks spreading to other areas. The extensive aerial survey would likewise segregate areas for treatment, or not, on the basis of levels of damage.

In the next phase, specific prescriptions are made for the treatment of various stands or groups of stands. For example, some areas may be scheduled for clearcutting, others for thinnings, and still others for pheromone trapping. Finally, the operation must be organized by setting treatment priorities and schedules and by allocating men and machinery to the tasks. We would generally schedule high-risk stands or areas suffering damage for treatment first, but this approach may be modified by constraints on accessibility, road building, and needs to space harvest cuts. The manager may also want to make contingency plans in the event that the forest suffers from serious disturbances. For example, suppression tactics (pheromone trapping) might be employed after windstorms or droughts in an attempt to maintain the beetle population below the outbreak threshold.

11.2. THE DOUGLAS-FIR TUSSOCK MOTH IN WESTERN NORTH AMERICA

The Douglas-fir tussock moth, *Orgyia pseudotsugata* (McD.) (Lepidoptera: Lymantriidae), is a native defoliator of true firs (*Abies* sp.) and Douglas-fir (*Pseudotsuga menziesii*) in western North America. The insect overwinters in the egg stage (egg diapause) and the larvae hatch in early spring, soon after the new shoots of its hosts start to elongate (Fig. 11.6a). The young larvae feed along the underside of the tender new needles consuming the mesophyll and leaving the upper cuticle intact (Fig. 11.6b). These needles soon dry and turn brown, giving the tree a scorched appearance. Fresh new foliage is essential for the young larvae and, if none is available, they will lower themselves on silk threads to the lower foliage or will be blown by the wind to adjacent trees. Young larvae may suffer very high mortality during outbreaks because the new foliage is rapidly exhausted. As larvae grow, they are able to consume the older foliage of their hosts and may even feed on nonhost trees when their preferred hosts are completely defoliated. However, survival of the larvae is reduced under these conditions, and those females that do survive lay few eggs.

Mature Douglas-fir tussock moth larvae are striking caterpillars with bright yellow, brown, and red markings and distinctive dorsal and terminal tufts of hair (Fig. 11.6c). In August the mature larvae begin to spin cocoons on the underside of foliage and branches, in holes and crevaces, or on the trunks of trees. They then pupate in these dense, hairy cocoons (Fig. 11.6d). Adult moths emerge from the cocoons in late August and September (Fig. 11.6c). The females are completely wingless and usually remain attached to their cocoons all their lives. Female tussock moths produce a powerful sex pheromone that guides the winged male to her for mating (the pheromone has been identified and synthesized and is available commercially). After mating, she lays her eggs in a hairy mass on her cocoon, where they remain over winter.

11.2.1. Interactions with the Tree

During outbreaks, Douglas-fir tussock moth populations can have severe impacts on their host plants. Some trees can be completely stripped of foliage for 1 year and sometimes 2 years. Many of these trees will die, particularly if they are already under stress from old age, overcrowding, poor site conditions, drought, or disease. In addition, a certain proportion of the trees that survive the

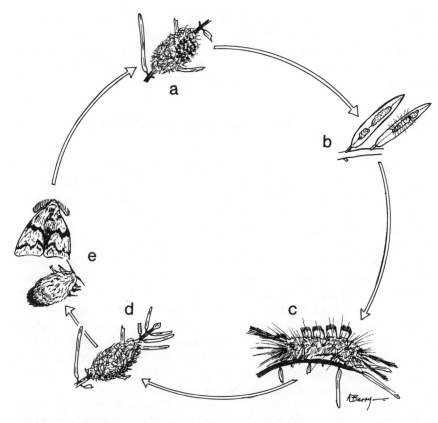

FIGURE 11.6. Life cycle of the Douglas-fir tussock moth. (a) Eggs (30–300/female) on female cocoons overwinter in diapause. (b) ln early spring, eggs hatch and first instar larvae feed along midribs of newly flushed needles. (c) The fifth instar larva is fully developed by midsummer. (d) Pupation occurs in dense, hairy cocoons. (e) Winged males and wingless females emerge and mate in autumn.

outbreak are killed by bark beetles before they are able to recover from the stress of heavy defoliation. However, although the forest is often thinned during tussock moth outbreaks, and groups of trees or large patches (up to 10 hectares) are sometimes killed, total mortality rarely exceeds 25–30% even in very severe outbreaks. Within 2 years after Douglas-fir tussock moth outbreaks have subsided, the forest as a whole looks green again.

Trees that survive heavy tussock moth defoliation may suffer other temporary or permanent injuries. Because the larvae concentrate their feeding in the upper crown, the top of the tree is sometimes killed, even though the tree as a whole survives. Tussock moth defoliation obviously reduces the photosynthesizing surface area and therefore the amount of carbohydrates produced. This loss of energy is reflected in lower stem growth (see Chapter 10, Fig. 10.3), shorter shoot elongation, and lower needle production in the year after defoliation. In other words, heavy defoliation results in a smaller quantity of new foliage in the year after defoliation, but foliage production returns to normal soon afterwards.

It is not known whether Douglas-firs and true firs change in any other ways in response to heavy defoliation by the Douglas-fir tussock moth. However, in light of recent discoveries with other forest defoliators, we might suspect subtle changes in the chemical composition, toughness, or nutrient content of the needles (see Chapters 3 and 4). It is also clear that tussock moths suffer higher mortality rates in heavily defoliated areas, particularly from virus disease and starvation (Fig. 11.7). It is possible that stress from food shortage or changed needle quality may make the insects more vulnerable to virus disease.

Tussock moth feeding undoubtedly increases the amounts of nutrients in the soil. The ground beneath heavily defoliated stands is often covered with insect feces and dead bodies. This fertilization results in greatly accelerated growth rates in surviving trees (see Chapter 10, Fig. 10.3) and may also cause changes in the chemistry or physical nature of new needles, which lower their value as tussock moth food.

11.2.2. Interactions with Natural Enemies

Douglas-fir tussock moths are attacked by a large number of insect predators and parasitoids, including 35 species of parasitic wasps, seven parasitic flies, two predaceous bugs, two predaceous beetles, and two predaceous lacewings. They are also eaten by a number of predaceous spiders and insectivorous birds.

Tussock moths are also infected by nuclear polyhedrosis viruses, particularly when their populations are very dense (Fig. 11.7). In fact, virus epizootics invariably precede the collapse of outbreaks. The spread of virus particles by

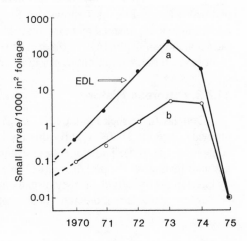

FIGURE 11.7. Changes in the density of Douglas-fir tussock moth larvae at Drumhill Ridge (a) and Sled Springs (b) in the Blue Mountains of Oregon. The outbreak collapsed in both places in 1974 due to starvation and virus disease. (After Mason and Luck, page 43 in Brookes *et al.*, 1978.) Economic damage level (EDL) is the density of larvae that causes visible defoliation.

dispersing larvae, parasites, and birds may also be involved in the collapse of less dense populations. For example, the two populations shown in Fig. 11.7 both collapsed to about the same level the same year, even though one of them never caused serious defoliation. Clearly, virus epizootics reduce the tussock moth population to very low levels over large areas and may thereby synchronize the next outbreak cycle.

Virus spray preparations can be produced in large quantities from laboratory reared tussock moths, and this biopesticide is registered in the United States for use against the Douglas-fir tussock moth. Because the virus is specific to this particular insect, it has no apparent effect on other organisms.

Despite considerable research on natural enemies of the tussock moth, their role in population regulation is still not clear. Predators and parasitoids cause considerable mortality to low populations and may sometimes cause the premature collapse of incipient outbreaks. Natural enemies may also play an important role in regulating tussock moth populations in regions in which outbreaks never occur, i.e., on cooler moister sites (see Section 11.2.3). However, the relationship between outbreaks and particular site and stand conditions suggests that the host plant has an important effect on tussock moth outbreaks. This hypothesis can be stated as follows: In areas in which outbreaks never occur, site conditions are normally favorable for the host plants and/or natural enemies, and tussock moth populations are limited to very low densities by a shortage of good-quality food and by interactions with natural enemies. For example, low-quality foliage of healthy hosts could reduce tussock moth reproduction and/or survival to such an extent that natural enemies are able to regulate population density. In areas in which outbreaks regularly occur, site conditions are generally unfavorable for the host plants and/or natural enemies and tussock moth food quality and survival is high. In these environments, tussock moth reproduction is too great for successful regulation by natural enemies, and the moth populations increase exponentially from very low densities initiated by virus epizootics in previous outbreaks. (*Note:* Fig. 11.7 shows that the growth rate before peak densities is logarithmic or exponential and that tussock moths in plot a have intrinsic growth rates higher than in plot b; $r_a = 0.83$; $r_b = 0.52$.) We will discuss the site factors involved in tussock moth outbreaks in more detail in the next section.

11.2.3. Outbreak History

Douglas-fir tussock moth outbreaks have occurred at regular 9- or 10-year intervals since the 1930s. Clearly this insect exhibits regular population cycles (Chapter 4, Fig. 4.8). The sequence of events is demonstrated by the record from northern Idaho over the past 40 years (Table 11.1). Outbreaks are often heralded by infestations of individual trees, usually around private homes or farms—Is it because these trees are more noticeable? Outbreaks may then develop over large

TABLE 11.1
History of Douglas-Fir Tussock Moth Outbreaks in North Idaho[a]

Year	Report
1944	Individual trees defoliated around farms and towns
1945	320 acres infested near Viola
1946	447,000 acres infested in Benwah, Latah, and Clearwater Counties
1947[b]	413,469 acres sprayed with DDT; outbreaks collapsed in sprayed and unsprayed areas due to virus epizootics
1955	35,000 acres infested near Orifino
1956[b]	Outbreak collapsed from natural causes (parasitoids or virus)
1961	Individual trees defoliated around farms and towns
1964	70,000 acres defoliated in Benewah and Latah Counties
1965[b]	225,000 acres infested; about one-half sprayed with DDT; outbreak collapsed due to virus epizootic and parasitoids
1971	Individual trees defoliated around homes in Coeur d'Alene
1972	Egg masses found in St. Joe National Forest but no defoliation seen
1973	98,900 acres defoliated in northern Idaho
1974[b]	115,000 acres defoliated in northern Idaho; 75,300 sprayed with DDT and outbreaks collapsed in both sprayed and unsprayed areas due to virus epizootics
1982	Individual trees infested near Genesee and increased moth captures in pheromone traps, but no visible defoliation seen in the forest
1983	Increasing moth captures in pheromone traps, but no visible defoliation evident in the forest
1984	Increasing moth captures in pheromone traps, but no visible defoliation

[a]From U.S. Forest Service Survey Reports and personal communication with William M. Ciesla, U.S.F.S.
[b]Year of collapse.

areas, as in 1946–1947, 1965, and 1973–1974, or may be restricted to small areas, as in 1955–1956. We might be tempted to speculate that variations in weather are responsible for reducing the extent of the outbreaks on susceptible sites. Cooler and wetter years, for example, would reduce moisture stress on the host and could thus decrease the nutritional quality of its foliage. These weather conditions would also reduce the rate of development of the insect and, perhaps, render it more vulnerable to natural enemy attack. If so, we can hypothesize that the early 1950s and 1980s were cooler and wetter than usual, while the early 1940s, 1960s and 1970s were normal or warmer and drier than usual.

Douglas-fir tussock moth outbreaks do not seem to spread but to build up more or less synchronously in particular areas. Outbreaks also seem to recur on particular sites, whereas elsewhere they never occur, even though tussock moths are present in low numbers. For example, outbreaks have never been recorded in the wet Douglas-fir and grand fir stands on the Pacific coast. Nor do they occur in wetter and cooler grand fir, white fir, or Douglas-fir stands in the interior. Instead, outbreaks seem to be associated with warm, dry sites at lower elevations on exposed ridgetops and sunny, south-facing, slopes. Defoliation is also more intense in dense, old, multistoried stands, in stands composed of a high percentage of the preferred hosts, and in stands growing on soils with poor water-holding capacities (thin volcanic ash deposits). The tussock moth can therefore be classified as exhibiting cyclical gradient outbreaks.

It is interesting, and probably significant, that outbreaks in northern Idaho are largely restricted to grand fir stands, whereas at lower elevations in eastern Washington, Oregon, and Idaho they are most severe in Douglas-fir stands. It has been noted that the low-elevation outbreaks are common in areas that were initially stocked with ponderosa pine, a more xeric species than Douglas-fir, or in the ponderosa pine–Douglas-fir transition zone (Fig. 11.8). On these sites, Douglas-fir is exposed to greater moisture stress and may therefore provide a more nutritious food supply for the tussock moth.

The establishment of Douglas-fir and many white fir stands on ponderosa pine sites is largely the result of human action. First, valuable low-elevation ponderosa pine stands were preferentially harvested during the late nineteenth century. Second, fire, which normally limited Douglas-fir and white fir establishment on these sites, has been controlled effectively since the early twentieth century. The result is that Douglas-fir and white fir now occupy sites that are too

FIGURE 11.8. Locations of major Douglas-fir tussock moth outbreaks in Washington and Oregon. Most outbreaks have occurred in fir stands growing on ponderosa pine sites or at the transition zone between fir and pine zones. (Redrawn after Martin and Williams, as reported in Brookes *et al.*, 1978, p. 58.)

warm and dry for optimal vigor and it is on these sites that tussock moth outbreaks usually occur.

A similar situation seems to have occurred in northern Idaho where low-elevation Douglas-fir was harvested in preference to grand fir. These sites were then occupied by seedlings from the remaining grand fir but, being drier and warmer, were not optimal for grand fir vigor. Thus outbreaks in northern Idaho are invariably associated with grand fir stands growing on warm, dry sites that were probably occupied originally by Douglas-fir. If these interpretations prove correct, Douglas-fir tussock moth outbreaks were probably rare events, or perhaps never occurred, before human incursion into the primeval forests of the inland Pacific Northwest.

11.2.4. Conventional Management Practices

In the past almost every serious Douglas-fir tussock moth outbreak has been sprayed from the air with the pesticide DDT (Table 11.1). Almost without exception, these spray projects were conducted in years when the moth population was collapsing, even in unsprayed areas, from natural causes (usually virus epizootics and starvation). The spray projects were justified in order to save foliage that would have been destroyed before the collapse.

In 1972 the U. S. Environmental Protection Agency (EPA) banned the use of DDT because of its harmful effects on fish, wildlife, and human health. Thus, when the expected tussock moth outbreaks occurred in Oregon, Washington and Idaho in that year, there was no pesticide available for use against the insect. The U.S. Forest Service, however, petitioned the EPA for special permission to use DDT, and this was granted in 1974. Aerial spraying was carried out on 421,000 acres in Oregon, Washington, and Idaho in that year, and the moth population collapsed in both sprayed and unsprayed areas.

Today, several insecticides are registered for use against the Douglas-fir tussock moth, including the synthetic carbamate carbaryl and a nuclear polyhedrosis virus formulation. The pheromone is also registered for operational use.

11.2.5. Practice for the Pest Manager

Following the procedures outlined in the previous section, the student should identify the structure and principal interactions of the Douglas-fir tussock moth system and should draw a diagram similar to that in Fig. 11.3. The student should then attempt to deduce the form of the density-dependent birth and death rate curves (e.g., as in Fig. 11.4). The next job is to outline a procedure for diagnosing tussock moth outbreaks, making prognoses of stand risk and outbreak behavior and prescribing preventative or therapeutic treatments. Finally, the student should attempt to develop a comprehensive pest management plan for the tussock moth. (Comments and some hints are provided in the answers to exercises.)

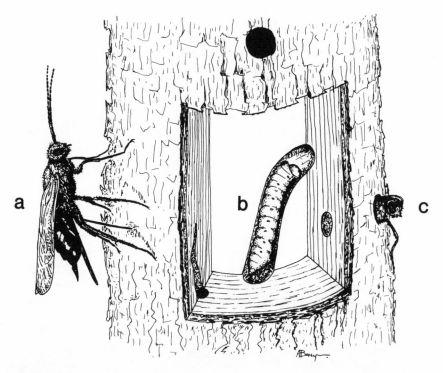

FIGURE 11.9. Life cycle of the European woodwasp. (a) In summer, the female deposits eggs, fungal spores, and toxic mucus in sapwood. (b) Larvae bore through sapwood feeding on fungal mycelia and pupate just below the bark surface the following spring. (c) In summer, the adults emerge to search for new host trees. Some larvae take longer to develop and may remain in the tree for 2–3 years.

11.3. THE EUROPEAN WOODWASP IN AUSTRALIA

The woodwasp *Sirex noctilio* F. (Hymenoptera: Siricidae) is a native of European pine forests. It is particularly prevalent around the Mediterranean area, where the large female wasps are often seen probing the boles of dead or dying pines with their long ovipositors (Fig. 11.9). The female lays her eggs deep within the wood; after hatching, the larvae bore tunnels through the sapwood. Although *Sirex noctilio* females occasionally attack living trees, and despite the fact that their larvae sometimes cause economic degrade of wood products, they are not considered serious pests in their native environment. For this reason, little research had been conducted on the ecology of siricid woodwasps before their accidental introduction into Australasia.

11.3.1. Outbreak and Management History

Sirex noctilio was first discovered in plantations of exotic pines on New Zealand's North Island during the early 1900s, presumably arriving there in pine-log imports from Europe. By 1927, it was causing serious mortality in exotic pine plantations, and efforts were initiated to introduce parasites for its control (biological control is one of the first thoughts that invariably spring to mind when confronted by an alien pest). Then, in 1952 a woodwasp infestation was discovered in a pine plantation near Hobart, Tasmania, and unsuccessful attempts were made to eradicate it by cutting and treating all infested trees (eradication is another "first thought" when confronted by restricted local infestations of an exotic pest). The woodwasp was next discovered in the State of Victoria, on the Australian mainland. At this point it posed a severe threat to the extensive pine plantations of western Victoria and neighboring South Australia.

Following this discovery, and in recognition of the threat to an expanding softwood industry, a coordinated research program was initiated to (1) search for and establish additional exotic natural enemies, (2) investigate the role of symbiotic fungi carried by the woodwasp, (3) determine whether host trees could resist attack, (4) study the biology and ecology of the woodwasp and its parasites, and (5) develop other methods of control. This coordinated attack by a large group of scientists from several institutions was one of the first and most successful cooperative research programs carried out on a forest insect pest.

11.3.2. Interactions with the Tree

The first and most critical interaction between the woodwasp and its coniferous host occurs during the act of oviposition. At this time the female *Sirex* commits her offspring to an uncertain environment that may or may not be suitable for development. We should not be surprised to learn that the act of oviposition is not a haphazard process in *Sirex noctilio.*

First, female woodwasps seem to be attracted to trees that are under severe physiological stress. Apparently the female can detect terpene vapors emanating from areas of the tree trunk that have low osmotic pressure and greatly reduced growth rates. Second, *Sirex* females apparently assess the suitability of the wood for their larvae before depositing an egg in the tree. If the host tissue is too moist (unsuitable), the female does not lay an egg but instead extrudes some toxic mucus into the hole she has bored—she "stings" the tree, if you like. She also deposits spores of a basidiomycete fungus (*Amylostereum areolatum*). Here, it turns out, is the key to the success achieved by *Sirex notilio* in Australasia.

The major plantation conifer in Australia and New Zealand is Monterey pine, *Pinus radiata,* a native of coastal California. This tree, usually called radiata pine in the Southern Hemisphere, grows remarkably well in its new home but is unfortunately rather sensitive to the toxic mucus of *Sirex noctilio*. Within 2 weeks of injection of the poison into the tree, either by woodwasp females or by artificial means, the needles in the upper crown turn chlorotic (yellowish), and stem growth is dramatically reduced. Although European pines may experience similar symptoms, they are generally not as intense or extensive as in radiata pine.

While the tree is reacting (in a state of shock?) to the "sting" of the woodwasp, the pathogenic basidiomycete begins to penetrate the xylem tracheids and rays, causing them to dry out. Embolisms of the tracheids break the water columns in the sapwood and the crown of the tree, cut off from its water supply, turns red and dies.

Healthy pines, however, can defend themselves against the fungus by producing resinous materials in the ray parenchyma. If these resins and polyphenols are formed rapidly enough, fungal growth is usually arrested and contained (i.e., the typical hypersensitive defense response discussed in Chapter 3). Although healthy Monterey pines are usually successful in defending themselves in this manner, they do not seem to have as effective a defense against *A. areolatum* as do their European relaltives. That *Pinus radiata* is more sensitive to the toxic secretions of *Sirex noctilio* and its fungal symbiont, is not surprising as these species have not had the evolutionary time to adapt to each other (Chapter 3).

Even though woodwasp females may actively seek out weakened hosts, they will undoubtably probe healthy trees when weak trees are unavailable. Continuous probing and deposition of toxins and fungi may weaken healthy pines to such an extent that they eventually succumb to repeated attack.

Let us now return to our probing female. If she finds that the osmotic pressure of the sapwood is suitable, she usually makes several drill holes at different angles from the insertion point; into these she places fungus spores with mucus and, in a separate hole, a single egg. *Sirex noctilio* eggs only hatch after fungal hyphae have grown out of adjacent drill holes and have penetrated the hole bearing the egg. After hatching, the larvae bore through the sapwood of the tree but feed exclusively on the fungae growing therein (Fig. 11.9). Larval development depends on the availability of fungus, and this is itself dependent on the moisture content of the sapwood. Larger and more fecund wasps are produced from trees that maintain their moisture content, hence promote fungal growth for a longer period of time.

When the density of larvae feeding in the tree is high, intense competition may occur, resulting in lower survival and smaller less fecund females. Obviously, very high *Sirex* populations will also deplete the total host population in a given region, and so reduce the food available for future generations.

It is apparent that physiological conditions of the host tree have important effects on *Sirex* reproductive success. Trees under stress from wind damage, drought, defoliation, nutrient deficiencies, and crowding are much less capable of defending themselves against the onslaught of insect and fungus. In fact, most

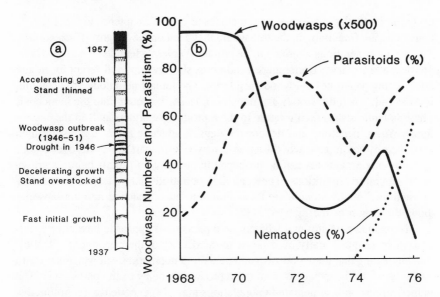

FIGURE 11.10. (a) Increment core from a Monterey pine that survived the 1946–1951 woodwasp outbreak in New Zealand and then showed growth acceleration after the stand was thinned from 600 to 130 trees per acre. (Redrawn from Gilmour, 1965.) (b) Reduction of the woodwasp population at Pittwater, Tasmania, after the introduction of insect parasitoids, mainly ichneumonids (- - -), and nematodes (. . .). (After Taylor, 1981.)

Sirex outbreaks seem to have originated in pine stands growing on poor sites or in overdense stands in need of thinning (Fig. 11.10a).

11.3.3. Interactions with Natural Enemies

Since the early 1930s a number of insect parasitoids have been introduced into New Zealand and Australia in an attempt to control the European woodwasp. The most important species, however, are two ibalids, *Ibalia leucospoides* and *I. rufipes* (Hymenoptera: Ibaliidae), which attack the eggs and early larval instars, and two ichneumonids, *Rhyssa persuasoria* and *Megarhyssa nortoni* (Hymenoptera: Ichneumonidae), which attack the larger larvae. These two parasitoid "guilds" are complementary as they attack the host at different times during its development.

In addition to the insect parasitoids, woodwasp populations in New Zealand were discovered to be parasitized by nematodes of the genus *Deladenus* (Nematoda: Neotylenchidae), which were apparently introduced accidentally in parasitoid shipments. After intensive studies, one of these nematode species (*D. siricidicola*) was chosen for introduction into Australia. A number of strains of this species were screened before a suitable one was selected for release, one that caused sterility in female woodwasps. The nematode parasite has a complex life cycle similar to that of bark beetle nematodes (e.g., see Chapter 5, Fig. 5.12). The nematodes are free living in the tree where they feed on the fungus, *A. areolatum*. Free-living generations continue to be produced unless a female nematode develops near a *Sirex* larva. Under these conditions, she develops a different form, and after mating penetrates the woodwasp larva. The female nematode grows within its host and, when the woodwasp pupates, releases its young into the hemocoel. These juvenile stages then migrate to the reproductive organs and, in the case of female *Sirex,* penetrate and devour the eggs. Nematode infected eggs are then deposited by the female woodwasp in a new tree to start the cycle over again.

Nematode parasites can be produced in quantity on fungus cultures grown under laboratory conditions. They can then be introduced quite easily into trees infested by *Sirex* by inoculating the culture into holes punched into the sapwood along the length of the infested hole.

Nematode parasites and ichneumonid parasitoids appear to have the capacity to increase numerically in response to the density of their prey (Fig. 11.10b). It is interesting that the woodwasp and parasitoid populations show a tendency to cycle, as would be expected from the reproductive delay of the parasitoids. The numerical response of nematode populations may also experience a reproductive delay.

11.3.4. Practice for the Pest Manager

Woodwasp populations in Australia seem to have generally subsided to low and relatively stable densities in recent years. Occasional outbreaks, however, have occurred in Victoria (1975–1979), northern Tasmania (1980–1983), and the Australian Capital Territory (1984). What is the future of *Sirex noctilio?* Does it still threaten Australian pine plantations? These questions can only be answered by analyzing the structure and potential dynamic behavior of *Sirex* populations. After this has been done, the student should be able to outline diagnostic and prescriptive procedures and to design a management plan (see answers to exercises for comments and hints).

11.4. ADDITIONAL MANAGEMENT PROBLEMS

Below are listed a series of forest insects that have been studied in sufficient detail to permit an analysis of population behavior and management strategies. I have used this list to assign special projects to my students and, for this purpose, some references are given to provide a place for them to begin.

REFERENCES AND SELECTED READINGS

The Spruce Beetle, *Ips typographus*

Bakke, A., 1982, The utilization of aggregation pheromone for the control of the spruce bark beetle, in: *Insect Pheromone Technology: Chemistry and Applications* (B. A. Leonhardt and M.

Beroza, eds.), pp. 219–229, American Chemical Society Symposium Series No. 190, Washington, D.C.

Bakke, A., 1983, Host tree and bark beetle interaction during a mass outbreak of *Ips typographus* in Norway, *Z. Angew. Entomol.* **96**:118–125.

Christiansen, E., 1985, *Ips/Ceratocystis*—infection of Norway Spruce: What is a deadly dosage?, *Z. Angew. Entomol.* **99**:6–11.

Christiansen, E., and Horntvedt, R., 1985, Combined *Ips/Ceratocystis* attack on Norway spruce, and defensive mechanisms of the trees, *Z. Angew. Entomol.* **96**:110–118.

Ravn, H. P., 1985, Expansion of the populations of *Ips typographus* (L.) (Coleoptera: Scolytidae) and their local dispersal following gale disaster in Denmark, *Z. Angew. Entomol.* **99**:26–33.

Thalenhorst, W., 1958, Grundzüge der Populationsdynamik des grossen Fichtenborkenkäfers *Ips typographus* L., *Schrift. Fotst. Fak. Univ. Gottingen* **21**:1–26. (with English summary)

Worrell, R., 1983, Damage by the spruce bark beetle in South Norway. 1970–80: A survey, and factors affecting its occurrence, *Rep. Norwegian For. Res. Inst.* **38(6)**:1–34.

The Douglas-fir Tussock Moth, *Orgyia pseudotsugata*

Berryman, A. A., 1978, Population cycles of the Douglas-fir tussock moth: The time-delay hypothesis, *Can. Entomol.* **110**:513–518.

Brookes, M. H., Stark, R. W., and Campbell, R. W., 1978, The Douglas-fir tussock moth: A synthesis, U.S. Forest Service Technical Bulletin 1585.

Mason, R. R., 1981, Host foliage in the susceptibility of forest sites in Central California to outbreaks of the Douglas-fir tussock moth, *Orgyia pseudotsugata* (Lepidoptera: Lymantriidae), *Can. Entomol.* **113**:325–332.

Mason, R. R., Torgerson, T. R., Wickman, B. E., and Paul, H. G., 1983, Natural regulation of a Douglas-fir tussock moth (Lepidoptera: Lymantriidae) population in the Sierra Nevada, *Environ. Entomol.* **12**:587–594.

Monserud, R. A., and Crookston, N. L., 1982, A user's guide to the combined stand prognosis and Douglas-fir tussock moth outbreak model, U.S. Forest Service, General Technical Report INT-27.

Stoszek, K. J., Mika, P. G., Moore, J. A., and Osborne, H. L., 1981, Relationships of Douglas-fir tussock moth defoliation to site and stand characteristics in Northern Idaho, *For. Sci.* **27**:431–442.

Wickman, B. E., 1978, A case study of a Douglas-fir tussock moth outbreak and stand conditions 10 years later, U.S. Forest Service Research Paper PNW-244.

Wickman, B. E., Mason, R. R., and Thompson, C. G., 1973, Major outbreaks of the Douglas-fir tussock moth in Oregon and California, U.S. Forest Service General Technical Report PNW-5. (good review of major outbreaks and control projects)

The European Woodwasp, *Sirex noctilio*

Bedding, R. A., and Akhurst, R. J., 1974, Use of the nematode *Deladenus siricidicola* in the biological control of *Sirex noctilio* in Australia, *J. Aust. Entomol. Soc.* **13**:129–135.

Coutts, M. P., 1969, The mechanism of pathogenicity of *Sirex noctilio* on *Pinus radiata*. I. Effects of the symbiotic fungus *Amylostereum* sp. (Thelophoraceae). II. Effects of *S. noctilio* mucus, *Aust. J. Biol. Sci.* **22**:915–924, 1153–1161.

Gilmour, J. W., 1965, The life cycle of the fungal symbiont of *Sirex noctilio*, *NZ J. For.* **10**:80–89.

Madden, J. L., 1968, Physiological aseects of host tree favorability for the woodwasp, *Sirex noctilio* F, *Nature (London)* **218**:189–190.

Madden, J. L., 1974, Oviposition behavior of the woodwasp *Sirex noctilio* F., *Aust. J. Zool.* **22**:341–351.

Madden, J. L., 1975, An analysis of an outbreak of the woodwasp, *Sirex noctilio* F. (Hymenoptera: Siricidae), in *Pinus radiata, Bull. Entomol. Res.* **65**:491–500.

Taylor, K. L., 1976, The introduction and establishment of insect parasitoids to control *Sirex noctilio* in Australia, *Entomophaga* **21**:429–440.

Taylor, K. L., 1978, Evaluation of the insect parasitoids of *Sirex noctilio* (Hymenoptera: Siricidae) in Tasmania, *Oecologia (Berl.)* **32**:1–10.

Taylor, K. L., 1980, Studies with *Sirex noctilio* (Hymenoptera: Siricidae) and its parasites that illustrate the importance of evaluating biological control attempts, *Acta Oecol. Appl.* **1**:181–187.

Taylor, K. L., 1983, The *Sirex* woodwasp: Ecology and control of an introduced forest insect, in: *The Ecology of Pests: Some Australian Case Histories* (R. L. Kitching and R. E. Jones, eds.), pp. 231–248, CSIRO, Melbourne, Australia.

The Spruce Budworm, *Choristoneura fumiferana*

Belyea, R. N., Miller, C. A., Baskerville, G. L., Kettela, E. G., Marshall, K. B., and Varty, I. W., 1975, The spruce budworm, *For. Chron.* **51**:1–26.

Clark, W. C., Jones, D. D., and Holling, C. S., 1979, Lessons for ecological policy design: A case study of ecosystem management, *Ecol. Model.* **7**:1–54.

Flexner, J. L., Bassett, J. R., Montgomery, B. A., Simmons, G. A., and Witter, J. A., 1983, Spruce–fir silviculture and the spruce budworm in the Lake States, Michigan Cooperative Forest Pest Management Program, CANUSA Handbook 83-2.

Hardy, Y. J., Lafond, A., and Hamel, L., 1983, The epidemiology of the current spruce budworm outbreak in Quebec, *For. Sci.* **29**:715–725.

Montgomery, B. A., Simmons, G. A., Witter, J. A., and Flexner, J. L., 1982, The spruce budworm handbook: A management guide for spruce–fir stands in the Lake States, Michigan Cooperative Forest Pest Management Program, CANUSA Handbook 82-7.

Morris, R. F. (ed.), 1963, The dynamics of epidemic spruce budworm populations, *Mem. Entomol. Soc. Can.* **31**:1–332. (especially sections 1, 2, 3, 18, 28, 30, 34, and 37)

Schmitt, D. M., Grimble, D. G., and Searcy, J. L. (eds.), 1984, Managing the spruce budworm in eastern North America, U.S. Forest Service, Agriculture Handbook No. 620.

Wellington, W. G., Fettes, J. J., Turner, K. B., and Belyea, R. M., 1950, Physical and biological indicators of the development of outbreaks of the spruce budworm, *Choristoneura fumiferana* (Clem.) (Lepidoptera: Tortricidae), *Can. J. Res.* **28**:308–331.

The Western Budworm, *Choristoneura occidentalis*

Campbell, R. W., Beckwith, R. C., and Torgerson, T. R., 1983, Numerical behavior of some western spruce budworm (Lepidoptera: Tortricidae) populations in Washington and Idaho, *Environ. Entomol.* **12**:1360–1366.

Campbell, R. W., Torgerson, T. R., and Srivastava, N., 1983, A suggested role for predaceous birds and ants in the population dynamics of the western spruce budworm, *For. Sci.* **29**:779–790.

Carlson, C. E., Fellin, D. G., and Schmidt, W. C., 1983, The western spruce budworm in northern Rocky Mountain forests: A review of ecology, insecticidal treatments and silvicultural practices, in: Management of Second-Growth Forests: The State of Knowledge and Research Needs (J. O'Loughlin and R. D. Pfister, eds.), pp. 76–103, Montana Forest and Conservation Experiment Station, School of Forestry, University of Montana, Missoula.

Fellin, D. G., 1983, Chemical insecticide vs. the western spruce budworm: After three decades, what's the score? *West. Wild.* **9**(1):8–12. (ecology, impacts and chemical and silvicultural control)

Fellin, D. G., Shearer, R. C., and Carlson, C. E., 1983, Western spruce budworm in northern Rocky Mountains, *West. Wild.* **9**(1):2-7. (ecology, impacts and chemical and silvicultural control)

Garton, E. O., and Langelier, L. A., 1985, Effects of stand characteristics on avian predators of western spruce budworm, in: The Role of the Host in the Population Dynamics of Forest Insects, Proceedings of the IUFRO Conference, Banff, Alberta (L. Safranyik, ed.), pp. 65-72, Pacific Forest Research Center, Victoria, British Columbia.

Harris, J. W. E., Alfaro, R. I., Dawson, A. F., and Brown, R. G., 1985, The western spruce budworm in British Columbia 1909-1983, Canadian Forestry Service Information Report BC-X-257.

Johnson, P. C., and Denton, R. E., 1975, Outbreaks of the western spruce budworm in the American northern Rocky Mountain Area from 1922 through 1971, U.S. Forest Service, General Technical Report INT-20.

Kemp, W. P., Everson, D. O., and Wellington, W. G., 1985, Regional climatic patterns and western spruce budworm outbreaks, U.S. Forest Service Technical Bulletin No. 1693.

Schmidt, W. C., Fellin, D. G., and Carlson, C. E., 1983, Alternatives to chemical insecticides in budworm-susceptible forests, *West. Wild.* **9**(1):13-19. (ecology, impacts and chemical and silvicultural control)

Thompson, A. J., 1979, Evaluation of key biological relationships of western budworm and its host trees, Canadian Forestry Service BC-X-186.

The Black-headed Budworm, *Acleris variana*

Miller, C. A., 1966, The black-headed budworm in eastern Canada, *Can. Entomol.* **98**:592-613.

Morris, R. F., 1959, Single factor analysis in population dynamics, *Ecology* **40**:580-588.

Silver, G. T., 1960, The relation of weather to population trends of the black-headed budworm, *Acleris variana* (Fern.) (Lepidoptera: Tortricidae), *Can. Entomol.* **92**:401-410.

The Gypsy Moth, *Lymantria dispar*

Campbell, R. W., 1975, The gypsy moth and its natural enemies, U.S. Forest Service, Agricultural Information Bulletin 381.

Campbell, R. W., 1979, Gypsy moth: Forest influence, U.S. Forest Service, Agricultural Information Bulletin 423.

Campbell, R. W., and Sloan, R. J., 1977, Release of gypsy moth populations from innocuous levels, *Enviorn. Entomol.* **6**:323-330.

Campbell, R. W., and Sloan, R. J., 1978, Numerical bimodality among North American gypsy moth populations, *Environ. Entomol.* **7**:641-646.

Doane, C. C., and McManus, M. L. (eds.), 1981, The gypsy moth: Research towards integrated pest management, U.S. Forest Service, Technical Bulletin 1584.

The Western Pine-shoot Borer, *Eucosma sonomana*

Daterman, G. E., 1982, Control of western pine shoot borer damage by mating disruption—A reality, in: *Insect Suppression with Controlled Release Pheromone Systems* (A. F. Kydonieus, M. Beroza, and G. Zweig, eds.), pp. 155-163, CRC Press, Boca Raton, Florida.

Stevens, R. E., and Jennings, D. T., 1977, Western pine-shoot borer: A threat to intensive management of ponderosa pine in the Rocky Mountain area and southwest, U.S. Forest Service, General Technical Report RM-45.

Stoszek, K. J., 1973, Damage to ponderosa pine plantations by the western pine-shoot borer, *J. For.* **71**:701-705.

The European Pine Shoot Moth, *Rhyacionia buoliana*

Miller, W. E., 1967, The European pine shoot moth—Ecology and control in the Lake states, *For. Sci. Monog.* **14**:1–72.

Heikkenen, H. J., 1981, The influence of red pine site quality on damage by the European pine shoot moth, in: Hazard-Rating Systems in Forest Insect Pest Management, (R. L. Hedden, S. J. Barras, and J. E. Coster, eds.), U.S. Forest Service, General Technical Report WO-27, pp. 35–44.

The Larch Budmoth, *Zeiraphera diniana*

Baltensweiler, W., 1964, *Zeiraphera griseana* Hubner (Lepidoptera: Tortricidae) in the European Alps. A contribution to the problem of cycles, *Can. Entomol.* **96**:792–800.

Baltensweiler, W., 1970, The relevance of changes in the composition of larch budmoth populations for the dynamics of its numbers, in: Dynamics of Populations, Proceedings of the Advanced Institute on the Dynamics of Numbers in Populations, Oosterbeek, Netherlands (P. J. den Boer and G. R. Gradwell, eds.), pp. 208–219, Centre for Agricultural Publications and Documentation, Wageningen, Netherlands.

Baltensweiler, W., Benz, G., Bovey, P., and Delucchi, V., 1977, Dynamics of larch budmoth populations, *Annu. Rev. Entomol.* **22**:79–100.

The Winter Moth, *Operophthera brumata*

Embree, D. G., 1966, The role of introduced parasites in the control of the winter moth in Nova Scotia, *Can. Entomol.* **98**:1159–1167.

Feeny, P., 1970, Seasonal changes in oak leaf tannins and nutrients as a cause of spring feeding by winter moth caterpillars, *Ecology* **51**:565–581.

Varley, C. G., Gradwell, G. R., and Hassell, M. P., 1973, *Insect Population Ecology: An Analytical Approach*. Blackwell, Oxford. (especially Chapter 7)

The Autumnal Moth, *Epirrita antumnata*

Haukioja, E., 1980, On the role of plant defenses in the fluctuation of herbivore populations, *Oikos* **35**:202–213.

Haukioja, E., 1982, Inducible defences of white birch to a geometrid defoliator, *Epirrita autumnata*, Proceedings of the 5th International Symposium on Insect–Plant Relationships, Wageningen, Netherlands.

Tenow, O., 1975, Topographical dependence of an outbreak of *Oporinia autumnata* Bkh. (Lep., Geometridae) in a mountain birch forest in northern Sweden, *Zoon* **3**:85–110.

The Forest Tent Caterpillar, *Malacosoma disstria*

Hildahl, V., and Reeks, W. A., 1960, Outbreaks of the forest tent caterpillar, *Malacosoma disstria* Hbn., and their effects on stands of trembling aspen in Manitoba and Saskatchewan, *Can. Entomol.* **92**:199–209.

Hodson, A. C., 1941, An ecological study of the forest tent caterpillar, *Malacosoma disstria* Hbn., in northern Minnesota. University of Minnesota, Agricultural Experiment Station Technical Bulletin 148.

Hodson, A. C., 1977, Some aspects of forest tent caterpillar population dynamics, in: Insect Ecology (H. M. Kulman and H. C. Chiang, eds.), pp. 5–16, University of Minnesota, Agricultural Experiment Station, Technical Bulletin 310.

Ives, W. G. H., 1973, Heat units and outbreaks of the forest tent caterpillar, *Malacosoma disstria* (Lepidoptera: Lasiocampidae). *Can. Entomol.* **105**:529–543.
Sippell, W. L., 1962, Outbreaks of the forest tent caterpillar, *Malacosoma disstria* Hbn., a periodic defoliator of broad-leaved trees in Ontario. *Can. Entomol.* **94**:408–416.

The Western Tent Caterpillar, *Malacosoma californicum pluviale*

Thompson, W. A., Vertinsky, I. B., and Wellington, W. G., 1979, The dynamics of outbreaks: Further simulation experiments with the western tent caterpillar, *Res. Popul. Ecol.* **20**:188–200.
Thompson, W. A., Vertinsky, I. B., and Wellington, W. G., 1981, Intervening in pest outbreaks: Simulation experiments with the western tent caterpillar, *Res. Popul. Ecol.* **23**:27–38.
Wellington, W. G., 1964, Qualitative changes in populations in unstable environments, *Can. Entomol.* **96**:436–451.
Wellington, W. G., 1977, Returning the insect to insect ecology: Some consequences for pest management, *Environ. Entomol.* **6**:1–8.

The Lodgepole Needle Miner, *Coleotechnites starki*

Stark, R. W., 1959, Population dynamics of the lodgepole needle miner, *Recurvaria starkii* Freeman, in Canadian Rocky Mountain Parks, *Can. J. Zool.* **37**:917–943.
Stark, R. W., and Cook, J. A., 1957, The effects of defoliation by the lodgepole needle miner, *For. Sci.* **3**:376–396.

The Southern Pine Beetle, *Dendroctonus frontalis*

Belanger, R. P., and Malac, B. F., 1980, Silviculture can reduce losses from the southern pine beetle. U.S. Department of Agriculture Handbook No. 576.
Coulson, R. N., 1979, Population dynamics of bark beetles, *Annu. Rev. Entomol.* **24**:417–447.
Hodges, J. D., and Pickard, L. S., 1971, Lightning in the ecology of the southern pine beetle, *Dendroctonus frontalis* (Coleoptera: Scolytidae), *Can. Entomol.* **103**:44–51.
Payne, T. L., Kudon, L. H., Walsh, K. D., and Berisford, C. W., 1985, Influence of infestation density on suppression of *D. frontalis* infestations with attractants, *Z. Angew. Entomol.* **99**:39–43.
Schowalter, T. D., Coulson, R. N., and Crossley, D. A., 1981, Role of southern pine beetle and fire in maintenance of structure and function of the Southeastern coniferous forest, *Environ. Entomol.* **10**:821–825.
Thatcher, R. C., Searcy, J. L., Coster, J. E., and Hertel, G. D. (eds.), 1982, The southern pine beetle, U.S. Forest Service Technical Bulletin 1631.

The Western Pine Beetle, *Dendroctonus brevicomis*

Miller, J. M., and Keen, F. P., 1960, Biology and control of the western pine beetle, USDA Forest Service Miscellaneous Publication 800.
Stark, R. W., and Dahlsten, D. L. (eds.), 1970, Studies on the population dynamics of the western pine beetle, *Dendroctonus brevicomis* Le Conte, University of California, Division of Agriculture, Scientific Publication.
Wood, D. L., 1972, Selection and colonization of ponderosa pine by bark beetles, in: *Insect/Plant Relationships* (H. F. van Emden, ed.), pp. 101–117, Blackwell, Oxford.

The Douglas-fir Beetle, *Dendroctonus pseudotsugae*

Furniss, M. M., McGregor, M. D., Foiles, M. W., and Partridge, A. D., 1979, Chronology and characteristics of a Douglas-fir beetle outbreak in northern Idaho, U.S. Forest Service, General Technical Report INT-50.

Furniss, M. M., Young, J. W., McGregor, M. D., Livingston, R. L., and Hamel, D. R., 1977, Effectiveness of controlled-release formulations of MCH for preventing Douglas-fir beetle (Coleoptera: Scolytidae) infestations in felled trees, *Can. Entomol.* **109**:1063–1069.

Lejeune, R. R., McMullen, L. H., and Atkins, M. D., 1961, The influence of logging on Douglas-fir beetle populations, *For. Chron.* **37**:308–314.

Rudinsky, J. A., 1966, Host selection and invasion by the Douglas-fir beetle, *Dendroctonus pseudotsugae* Hopkins, in Coastal Douglas-fir forests, *Can. Entomol.* **98**:98–111.

Ryker, L. C., 1984, Acoustic and chemical signals in the life cycle of a beetle, *Sci. Am.* **250**(6):112–123.

The Mountain Pine Beetle, *Dendroctonus ponderosae*

Amman, G. D., and Baker, B. H., 1972, Mountain pine beetle influence on lodgepole pine stand structure, *J. For.* **70**:204–209.

Berryman, A. A., 1982, Mountain pine beetle outbreaks in Rocky Mountain lodgepole pine forests, *J. For.* **80**:410–413.

Berryman, A. A., Amman, G. D., Stark, R. W., and Kibbee, D. L. (eds.), 1978, Theory and practice of mountain pine beetle management in lodgepole pine forests, Proceedings of a Symposium at Washington State University, Forestry, Wildlife and Range Experiment Station, University of Idaho, Moscow.

Raffa, K. F., and Berryman, A. A., 1983, The role of host resistance in the colonization behavior and ecology of bark beetles (Coleoptera: Scolytidae), *Ecol. Monog.* **53**:27–49.

Safranyik, K., Shrimpton, D. M., and Whitney, H. S., 1974, Management of lodgepole pine to reduce losses from the mountain pine beetle, Canadian Forestry Service, Technical Report 1.

The Great Spruce Bark Beetle, *Dendroctonus micans*

Brown, J. M. B., and Bevan, D., 1966, *The Great Spruce Bark Beetle, Dendroctonus micans, in North West Europe,* Her Majesty's Stationary Office, London. (see references for a number of important papers, mostly in German)

Grégoire, J.-C., 1985, Host colonization strategies in *Dendroctonus:* Larval gregariousness or mass attack by adults? in: The Role of the Host in Population Dynamics of Forest Insects (L. Safranyik, ed.), pp. 147–154, Proceedings of the IUFRO Conference, Banff, Alberta, Pacific Forest Research Centre, Victoria, British Columbia.

Grégoire, J.-C., Merlin, J., Pasteels, J. M., Jaffuel, R., Vouland, G., and Schvester, D., 1985, Biocontrol of *Dendroctonus micans* by *Rhizophagus grandis* Gyll. (Col., Rhizophagidae) in the Massif Central (France), *Z. Angew. Entomol.* **99**:182–190.

Grégoire, J.-C., and Pasteels, J. M. (eds.), 1984, Biocontrol of bark beetles (*Dendroctonus micans*). Proceedings of a Symposium organized by the Commission of the European Communities and the Université Libre de Bruxelles, Brussels.

The Spruce Beetle, *Dendroctonus rufipennis*

Baker, B. H., and Kemperman, J. A., 1974, Spruce beetle effects on a white pine stand in Alaska, *J. For.* **72**:423–425.

Hard, J. S., Werner, R. A., and Holsten, E. H., 1983, Susceptibility of white spruce to attack by spruce beetles during the early years of an outbreak in Alaska, *Can. J. For. Res.* **13**:678–684.

Schmid, J. M., and Frye, R. H., 1976, Stand ratings for spruce beetle, U.S. Forest Service, Research Note RM-309.

Werner, R. A., Baker, B. H., and Rush, P. A., 1977, The spruce beetle in white spruce forests of Alaska, U.S. Forest Service, General Technical Report PNW-61.

Werner, R. A., and Holsten, E. H., 1983, Mortality of white spruce during a spruce beetle outbreak on the Kenai Peninsula in Alaska, *Can. J. For. Res.* **13**:96–101.

European Elm Beetle, *Scolytus scolytus*

Beaver, R. A., 1967, The regulation of population density in the bark beetle, *Solytus scolytus* (F.), *J. Anim. Ecol.* **36**:435–451.

Beaver, R. A., 1969, Natality, mortality and control of the elm bark beetle, *Scolytus scolytus* (F.) (Col., Scolytidae), *Bull. Entomol. Res.* **59**:537–540.

Blight, M. M., Wadhmas, L. J., and Wenham, M. J., 1979, Chemically mediated behavior in the large elm bark beetle, *Scolytus scolytus, Bull. Entomol. Soc. Am.* **25**:122–124.

Brasier, C. M., and Gibbs, J. N., 1973, Origin of the Dutch elm disease epidemic in Britain, *Nature (Lond.)* **242**:607–609.

Gibbs, J. N., 1978, International epidemiology of Dutch elm disease, *Annu. Rev. Phytopathol.* **16**:287–307.

Gibbs, J. N., 1978, Development of the Dutch elm disease epidemic in southern England, 1971–6. *Ann. Appl. Biol.* **88**:219–228.

Heybrock, H. M., Elgersma, D. M., and Scheffer, R. J., 1982, Dutch elm disease: An ecological accident, *Outlook Agric.* **11**:1–9.

The Fir Engraver, *Scolytus ventralis*

Berryman, A. A., 1969, Responses of *Abies grandis* to attack by *Scolytus ventralis* (Coleoptera: Scolytidae), *Can. Entomol.* **101**:1033–1041.

Berryman, A. A., 1973, Population dynamics of the fir engraver, *Scolytus ventralis* (Coleoptera: Scolytidae). 1. Analysis of population behavior and survival from 1964–71, *Can. Entomol.* **105**:1465–1488.

Bordasch, R. P., and Berryman, A. A., 1977, Host resistance to the fir engraver beetle, *Scolytus ventralis* (Coleoptera: Scolytidae). 2. Repellency of *Abies grandis* resins and some monoterpenes, *Can. Entomol.* **109**:95–100.

Ferrell, G. T., 1973, Weather, logging and tree growth associated with fir engraver attack scars in white fir, U.S. Forest Service Research Paper PSW-92.

Hertert, H. D., Miller, D. L., and Partridge, A. D., 1975, Interaction of bark beetles (Coleoptera: Scolytidae) and root-rot pathogens in grand fir in northern Idaho, *Can. Entomol.* **107**:899–904.

Struble, G. R., 1957, The fir engraver, a serious enemy of western true firs, U.S. Department of Agriculture, Product Research Department No. 11.

The Ambrosia Beetle, *Trypodendron lineatum*

Borden, J. H., and McLean, J. A., 1981, Pheromone-based suppression of ambrosia beetles in industrial timber processing areas, in: *Management of Insect Pests with Semiochemicals* (E. R. Mitchell, ed.), pp. 133–154, Plenum, New York.

Chapman, J. A., and Nijholt, W. W., 1980, Time of attack flight of ambrosia beetle, *Trypodendron lineatum* (Oliv.) (Coleoptera: Scolytidae) in relation to weather in Coastal British Columbia. Canadian Forestry Service, Pacific Forest Research Center, Victoria, BC-R-5.

McMullen, D. L., 1956, Ambrosia beetles and their control in British Columbia, *For. Chron.* **32:**31–43.

Moek, H. A., 1970, Ethanol as the primary attractant for the ambrosia beetle *Trypodendron lineatum* (Coleoptera: Scolytidae), *Can. Entomol.* **102:**985–995.

Prebble, M. L., and Graham, K., 1956, Studies of attack by ambrosia beetles in softwood logs on Vancouver Island, British Columbia, *For. Sci.* **3:**90–112.

The Japanese Pine Sawyer, *Monochamus alternatus*

Dropkin, V. H., Foudin, A., Kondo, E., Linit, M., Smith, M., and Robbins, K., 1981, Pinewood nematode: A threat, *Plant Dis.* **65:**1022–1027.

Kobayashi, F., Yamane, A., and Ikeda, T., 1984, The Japanese pine sawyer beetle as the vector of pine wilt disease, *Annu. Rev. Entomol.* **29:**115–135.

Mamiya, Y., 1972, Pine wood nematode, *Bursaphelenchus lignicolus* Mamiya and Kiyohara, as a causal agent of pine wilting disease, *Rev. Plant Protect. Res.* **5:**46–60.

Wingfield, M. J., Blanchette, R. A., Nicholls, T. H., and Robbins, K., 1982, The pine wood nematode: A comparison of the situation in the United States and Japan, *Can. J. For. Res.* **12:**71–75.

The Larch Sawfly, *Pristiphora erichsonii*

Graham, S. A., 1956, The larch sawfly in the Lake States, *For. Sci.* **2:**132–160.

Ives, W. G. H., 1976, The dynamics of larch sawfly (Hymenoptera: Tenthredinidae) populations in southeastern Manitoba, *Can. Entomol.* **108:**701–730.

Turnock, W. J., 1972, Geographical and historical variability in population patterns and life systems of the larch sawfly (Hymenoptera: Tenthredinidae), *Can. Entomol.* **104:**1883–1900.

The European Pine Sawfly, *Neodiprion sertifer*

Larsson, S., and Tenow, O., 1984, Areal distribution of a *Neodiprion sertifer* (Hym., Diprionidae) outbreak on Scots pine as related to stand condition, *Holarctic Ecol.* **7:**81–90.

Niemelä, P., Tuomi, J., Manila, R., and Ojala, P., 1984, The effect of previous damage on the quality of Scots pine foliage as food for Diprionid sawflies, *Z. Angew. Entomol.* **98:**33–43.

Lyons, L. A., 1977, On the population dynamics of *Neodiprion* sawflies, in: Insect ecology (H. M. Kulman and H. C. Chiang, eds.), pp. 48–55, Technical Bulletin 310, University of Minnesota, Agricultural Experiment Station.

Pschorn-Walcher, H., and Eichorn, O., 1973, Studies on the biology of egg parasites (Hym.: Chalcidoidea) of the pine sawfly, *Neodiprion sertifer* (Geoff.) (Hym.: Diprionidae) in Central Europe, *Z. Angew. Entomol.* **74:**286–318.

Jack Pine Sawfly, *Neodiprion swainei*

McLeod, J. M., 1966, The spatial distribution of cocoons of *Neodiprion swainei* in a jack pine stand. I. A chartographic analysis of cocoon distribution with reference to predation by small mammals, *Can. Entomol.* **98:**430–447.

McLeod, J. M., 1970, The epidemiology of the Swaine jack pine sawfly, *Neodiprion swainei* Midd, *For. Chron.* **46:**126–133.

McLeod, J. M., 1972, The Swaine jack pine sawfly, *Neodiprion swainei* life system; evaluating the long-term effects of insecticide applications in Quebec, *Environ. Entomol.* **1:**371–381.

McLeod, J. M., 1974, Bird population studies in the Swaine jack pine sawfly life system, Canadian Forestry Service, Information Report LAV-X-10.

The Beech Scale, *Cryptococcus fagisuga*

Houston, D. R., Parker, E. J., and Lonsdale, D., 1979, Beech bark disease: Patterns of spread and development of the initiating agent *Cryptococcus fagisuga*, *Can. J. For. Res.* **9**:336–344.

Houston, D. R., and Wainhouse, D. (eds.), 1983, Proceedings of the IUFRO Beech Bark Disease Working Party Conference, U.S. Forest Service, General Technical Report WO-37.

Wainhouse, D., and Howell, R. S., 1983, Intraspecific variation in beech scale populations and in susceptibility of their host *Fagus sylvatica*, *Ecol. Entomol.* **8**:351–359.

The Hemlock Scale, *Fiorinia externa*

McClure, M. S., 1977, Resurgence of the scale, *Fiorinia externa* (Homoptera: Diaspididae), on hemlock following insecticide application, *Environ. Entomol.* **6**:480–484.

McClure, M. S., 1979, Self-regulation in populations of the elongate hemlock scale, *Fiorinia externa* (Homoptera: Diaspididae), *Oecologia* **39**:25–36.

McClure, M. S., 1980, Foliar nitrogen: A basis for host suitability for elongate hemlock scale, *Fiorinia externa* (Homoptera: Diaspididae), *Ecology* **61**:72–79.

McClure, M. S., 1980, Competition between exotic species: Scale insects on hemlock, *Ecology* **61**:1391–1401.

The Balsam Woolly Adelgid, *Adelges piceae*

Arthur, F. H., and Hain, F. P., 1984, Seasonal history of the balsam wooly adelgid (Homoptera: Adelgidae) in natural stands and plantations of Fraser fir, *J. Econ. Entomol.* **77**:1154–1158.

Balch, R. E., 1952, Studies of the balsam wooly aphid, *Adelges piceae* (Ratz.) and its effect on balsam fir, *Abies balsamea* (L.) Mill, Canadian Department of Agriculture Publication No. 867.

Hain, F. P., Mawby, W. D., Cook, S. P., and Arthur, F. H., 1983, Host conifer reaction to stem invasion, *Z. Angew. Entomol.* **96**:247–256.

Hain, F. P., and Arthur, F. H., 1985, The role of atmospheric deposition in the latitudinal variation of Fraser fir mortality caused by the balsam woolly adelgid, *Adelges piceae* (Ratz.) (Hemipt., Adelgidae): A hypothesis, *Z. Angew. Entomol.* **99**:145–152.

Puritch, G. S., and Petty, J. A., 1971, Effect of balsam woolly aphid, *Adelges piceae* (Ratz.) infestation on the xylem of *Abies grandis* (Dougl.) Lindley, *J. Exp. Bot.* **22**:946–952.

The Textured Psyllid, *Cardiaspina albitextura*

Clark, L. R., 1964, Predation by birds in relation to the population density of *Cardiaspina albitextura* (Psyllidae), *Aust. J. Zool.* **12**:349–361.

Clark, L. R., 1964, The population dynamics of *Cardiaspina albitextura* (Psyllidae), *Aust. J. Zool.* **12**:362–380.

Clark, L. R., and Dallwitz, M. J., 1975, The life system of *Caridaspina albitextura* (Psyllidae), *Aust. J. Zool.* **23**:523–561.

Loyn, R. H., Runnalls, R. G., Forward, G. Y., and Tyers, J., 1983, Territorial bell miners and other birds affecting populations of insect prey, *Science* **221**:1411–1412.

The Walkingstick, *Didymuria violescens*

Readshaw, J. L., 1965, A theory of phasmatid outbreak release. *Aust. J. Zool.* **13**:475–490.

EPILOGUE

THE HUMAN IMPACT
Economic Desires versus Ecological Realities

This we know. The earth does not belong to man; man belongs to the earth. This we know. All things are connected. . . . Man did not weave the web of life; he is merely a strand in it. Whatever he does to the web, he does to himself.

Chief Seattle, 1854, excerpted from a documentary film produced by the Southern Baptist Convention's Radio and Television Commission. Copyright 1972, all rights reserved.

Needles on spruce trees in the Black Forest of Germany turn yellow and fall. Multitudes of fish expire in Scandinavian and Canadian lakes. Thousands of firs and spruces decline and die in the Appalachian Mountains of eastern North America, and scores of ponderosa pines succumb in the San Bernadino Mountains east of Los Angeles. All over the world forests are showing unprecedented declines in growth and productivity. *Waldsterben,* the German word for forest decline, is frequently heard at scientific meetings, public debates, and in the halls of justice and legislation.

Although still controversial, the evidence is becoming overwhelming that air pollution is the major factor contributing to *waldsterben.* The most common pollutants are oxides of carbon, sulfur, and nitrogen, along with ozone and various heavy metals such as lead, aluminum, copper, and zinc. These materials are produced by the burning of fossil fuels in automobile engines, power plants, and factories. Upon interaction with moisture in the atmosphere, the sulfur and nitrogen oxides produce sulfuric and nitric acids, which are deposited as "acid rain."

Pollution is not a phenomenon unique to *Homo sapiens* but is rather a natural consequence of overpopulation; i.e., large populations tend to foul their environments. As the noted economist Kenneth Boulding remarked in his address to the Ecological Society of America in 1984, oxygen was probably the first pollutant produced by organisms living on the planet earth. Because of this oxygen pollution, the anaerobic organisms that produced it were largely replaced by aerobic organisms that evolved to use the oxygen as a resource. This is the *law of succession* in action—which states that whenever a species population alters its environment to its disfavor, be it by pollution or any other effect, it will be succeeded by populations of other species that find the changed environment to their liking. We see this law in action in natural environments where pioneer species gradually disappear from a locality because their intolerant seedlings cannot grow in the shaded environment created by their parents (Chapter 3).

Pollutants have both direct and indirect influences on forest productivity. Direct effects are seen in the symptoms of crown die-back, where needles exposed to the sun turn chlorotic and then fall. However, more serious direct problems may be caused by soil acidification, particularly in soils with low pH levels. In these cases acidification may lead to the leaching of minerals, such as calcium and magnesium, which are important for tree growth and survival.

The indirect effects of pollutants may be even more serious in the long run. Soil acidification can influence microorganisms that play important roles in nutrient cycling and uptake. In addition, pollution induced stress can reduce tree resistance to insects and diseases. For example, extensive mortality caused by the balsam woolly adelgid in the Appalachian Mountains seems to be associated with pollution stress. During the past 50 years, outbreaks of forest insects have also increased dramatically in Europe, particularly in areas exposed to air pollutants and acid rain, and several new pest species have appeared on the scene. These complex influences on the stability of forest ecosystems are extremely difficult to predict. However, we do know that stress on forest trees can induce outbreaks of eruptive pests which can result in destruction of vast forested areas (Chapter 4).

Although air pollution undoubtedly takes its toll in forests close to centers of industry and population, the effects can be felt over large areas. Forest decline, for instance, is most severe on mountain tops where the trees may be bathed in a cloud of "chemical soup." Lakes are also affected in Scandinavia, far from the sources of pollution in England, Germany, and Poland. Indeed, we are beginning to see global effects as increasing levels of carbon dioxide and other pollutants accumulate in the atmosphere—e.g., the *greenhouse effect*.

Air pollution is perhaps the most spectacular and serious human impact on the world's forests, but it is certainly not the only one. The world-wide distribution of forest trees is being continuously changed as exotic species are used more and more in plantation forestry: Radiata pine, a California native, has become the major timber species in Australia and New Zealand, while sitka spruce, Douglas-fir, and lodgepole pine are being planted with increasing frequency in Europe. We should expect trouble from insects in these exotic plantations: Will native forest insects adapt to feed on these exotic trees that have not had the evolutionary time to evolve defensive reactions? In England, for example, an insect that normally feeds on deciduous trees, the winter moth, seems to have adapted to feed on sitka spruce, a conifer. Sitka spruce is also attacked by a native *Dendroctonus* bark beetle, whereas lodgepole pine suffers outbreaks of native pine beauty moths and larch budmoths.

On the other side, human commerce and transport systems are inadvertently spreading pests around the world at an increasing rate. The list seems endless, but some of the most disastrous introductions are the Dutch elm disease, chestnut blight, pine-wood nematode, European woodwasp, smaller European elm bark beetle, gypsy moth, and balsam woolly adelgid. Once again, the incompatibility of unadapted plant–herbivore species is the apparent cause of the problem. The result is an instability that often leads to widespread mortality of the host plant. As humans continue to move themselves, their products, and forest trees around the globe we should expect further problems from unadapted plant–herbivore interactions.

Even overlooking these problems as unavoidable or impossible to deal with, *Homo sapiens* has often been careless in husbanding forest resources. The approach to foresty has often been exploitation rather than husbandry. Thus, North American coniferous forests were plundered for the more valuable species, causing drastic changes in species compositions and stand structures, many of which are more susceptible to insect attack (see Section 11.2). Nowadays, exploitation forestry has spread to the tropics where forests are being harvested and converted to agricultural lands at an alarming rate. The problem is even more severe in the tropics because soils are less stable and torrential rains often cause massive erosion.

Many of the problems appear to have resulted from conflicts between economic desires and ecological realities. The forest economist is concerned with reaping the maximum profit from investments made in stand establishment and improvement. In other words, his thinking revolves around the growth rate and value of the tree species. However, if this thinking neglects the ecological realities, say by planting economically desirable species on suboptimal sites, the investment may be completely destroyed by insect outbreaks. In addition, the forest owner may face lawsuits if the outbreak spreads to adjacent ownerships (e.g., an eruptive outbreak). Even if timber values can be saved by suppressive tactics, the costs involved will eat into the profit margin. The economic–ecological conflict is captured nicely by a story I heard in Norway while viewing the great spruce beetle outbreak of the 1970s. Up until the mid-1900s, the bark was peeled from all spruce logs to prevent beetles from breeding under the bark. The practice was originally implemented after the great outbreak of the 1850s, but was abandoned for economic reasons about 100 years later. One can imagine the conversation between forester and entomologist:

FORESTER: Why are we going through this labor-intensive and expensive process of peeling spruce logs in the woods?

ENTOMOLOGIST: Why, to prevent spruce bark beetles from breeding under the bark, of course.

FORESTER: What's a spruce bark beetle?

ENTOMOLOGIST: It's a small black beetle with spines on the tail-end that bores galleries, sort of tuning-fork shaped, under the bark.

FORESTER: Hmm! Well I've never seen one of those and I've been working in the woods for 40 years.

ENTOMOLOGIST: That's because peeling the bark from spruce logs, so that the beetles have nowhere to breed, works!

Whatever the discussion, and the entomologists did have their say, the economic desire once again triumphed over the ecological reality. One question, however, was answered. Norwegian foresters no longer need entomologists to explain what a spruce bark beetle looks like. They have seen thousands of beetles captured in plastic pheromone traps and numerous galleries engraved on the 5 million cubic meters of dead spruces.

Is it possible to resolve these conflicts between economics and ecology? I believe it is, but only if we accept the fact that forests are ecological entities governed by natural laws—laws that cannot be changed by human actions and to which humans themselves are subject. In other words, we must listen to the wisdom of Chief Seattle and realize that we are part of the ecological web— whatever we do to the web we do unto ourselves. If we accept this fundamental notion, then it follows that economic desires must be subservient to the ecological realities and, under these conditions, the conflict disappears because economic decisions become subject to the laws of nature.

It will help if we can learn to value forests for things other than logs. What is the value of oxygen that forests produce and we breathe? What is the value of clear water and air cleansed by leaves and soil? Is there a value to sitting in a forest glade and watching a deer, or just knowing that one can if one wants to? Only if these values are acknowledged can we make fair decisions between logs and trees.

Finally, it will help if we can answer the ethical question of descendent rights—do one's children and one's children's children have a right to experience the forest as we know it today? If the answer to this question is yes, then we must manage the forest according to the laws of nature, for only in this way can · we guarantee their stability and persistence.

REFERENCES AND SELECTED READINGS

Baltensweiler, W., 1985, "Waldsterben": forest pests and air pollution, *Z. Angew. Entomol.* **99**:77–85.

Postel, S., 1984, Air pollution, acid rain, and the future of forests, *World Watch Paper 58.*

Sitwell, N., 1984, Our trees are dying, *Science Digest* **92**:39–48.

Smith, W. H., 1985, Forests and air quality, *J. For.* **83**:83–92.

Stark, R. W., and Cobb, F. W., 1969, Smog injury, root diseases and bark beetle damage in ponderosa pine, *Calif. Agric.* **23**:13–15.

ANSWERS TO EXERCISES

PART II: ECOLOGICAL EXERCISES

II.1 a. The diagram should have at least one positive and one negative feedback loop.
   b. See Fig. 11.3 (Chapter 11) for a population diagram for a similar bark beetle system.
   c. See Fig. A1a on page 254.
   d. A stable low-density equilibrium, E_1, an unstable threshold T and a cyclical high-density equilibrium, E_2 (Fig. A1a).
   e. See Fig. A1b. The outbreak threshold is zero in nonresistant stands (R_1), increases with stand resistance $(R_2 - R_4)$, and is infinite in very resistant stands (R_5).
   f. See Fig. A1c.
   g. Pulse eruptions (see Fig. 4.11a).
II.2 a. Use Fig. 4.1 as a model. Population density should have an effect on environmental favorability (escape from predators and food availability) and on the basic environment (killing host trees after several years heavy defoliation and the numbers of parasitoids and pathogens).
   b. Starting at a very low density, the environment becomes less favorable as populations grow because natural enemies notice and then concentrate their feeding on the moth. At intermediate densities the environment becomes more favorable because natural enemies are satiated. At high densities the environment becomes very unfavorable because food is exhausted and virus epizootics spread through the population.
   c. Birth rates should decline only at very high densities. Death rates should rise at first, then decline, and then rise again, giving a system with three potential equilibrium points. The low-density equilibrium is probably highly stable [rapid $(-)$ feedback], the second is an unstable threshold, the third is probably oscillatory or cyclic [delayed $(-)$ feedback]. Potentially a sustained eruptive type of outbreak behavior.
   d. Outbreaks if maximum birth rate is increased sufficiently, low-density behavior if it is reduced sufficiently.
   e. More frequent outbreaks along edges of forest (see Campbell and Sloan, 1977).
II.3 a. Equilibrium cone crop around 20,000 cones/acre, for beetles around 1600/acre. Beetle

253

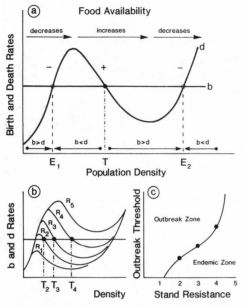

FIGURE A1. (a) Deduction of the individual death rate curve for a *Dendroctonus* beetle under the assumption that environmental favorability is determined by the food supply and that the birth rate (b) remains constant (allowing the birth rate to change with density does not affect the result; see Fig. 11.4): At first the death rate rises with increasing beetle density because the food available per individual decreases (negative feedback results when the quantity of food, severely weakened trees, remains constant while beetle density rises). As density gets higher beetles are able to overwhelm resistant trees, the food supply expands, and the death rate falls (positive feedback). At very high densities trees are killed faster than they can be replaced and the death rate rises again (delayed negative feedback). (b) If we repeat the exercise above with a less vigorous stand (R_1) the (+) feedback begins at a lower density because fewer beetles are needed to overcome the resistance of the average tree, while if the stand has greater resistance (R_5) the (+) feedback begins at much higher population density. (c) Threshold points from (b) plotted against stand resistance.

population peaks in 1970, 1972, and 1976 either follow or coincide with heavy cone crops in 1969, 1972, and 1975. Beetle populations decline to very low levels in the year following poor cone crops.

   b. A fast-acting negative feedback loop between beetle population density and environmental favorability; i.e., the environment becomes less favorable as the number of cones available per beetle decreases or the beetle population increases.

   c. Births + immigrations exceed deaths + emigrations when $r_t > 0$ or when $F_t > 12$ cones/beetle and vice-versa. According to the graph, equilibrium would occur at 12 cones/beetle, or $30,000/12 = 2500$ beetles per acre.

   d. Equilibrium beetle population density E is a linear function of cone crop size, $E = C/12$. Beetle populations grow or decline entirely in response to cone crop size, therefore pulse gradient outbreaks occur.

II.4 a. A graph of prey caught per predator per day on prey density shows that predator #1 has a cyrtoid functional response probably caused by the capture rate declining with prey density because more and more time is spent in subduing, eating and digesting the prey rather than in searching and capture activities. Predator #2 has a sigmoid functional response probably caused by an increasing capture rate at relatively low prey density, as predators switch to this prey when there is increased abundance. At higher prey densities, however, the capture rate drops off for the same reasons as above.

   b. The individual prey death rate due to a single individual of predator #1 declines with prey density and crosses b, which is constant with prey density at $N = 280$. This is an unstable equilibrium point because the prey declines to extinction if its density is <280, while it grows continuously if $N > 280$. This predator cannot regulate the prey population.

    c. No. The birth rate always exceeds the death rate and so the prey population will grow continuously.

    d. Yes, at a density of about 240 prey. An outbreak threshold occurs at a density of 600 prey.

PART III: MANAGEMENT EXERCISES

III.1.   *Stand 1:* Decreasing at 80% confidence level.
        Stand 2: Increasing at 90% confidence level.
        Stand 3: Undecided; continue sampling up to 30 more trees.

III.2 a. $I = 4.76$ = increasing trend because $I > 1$. The population will increase if I remains constant, a shaky assumption.

     b. Decreasing, $I = 0.75$. Probably more reliable than the trend index because parasites are included in the prediction.

     c. $I = 0.42$ = decreasing because $I < 1$.

III.3.   PGR = 0.48; RSI = 0.03

III.4.   RV = 3 + 2 + 3 + 3 = 11 = high risk

III.5.   Penalty score = 0 + 5 + 0 + 0 = 5 = moderate risk (categorie III)

III.6. $X' = 9.643 + 0.032 \times 0 - 0.078 \times 2 - 0.047 \times 30 - 1.419 \times 2 = 5.24$
       $P = 1/[1 + \exp(-5.24)] = 1/1.01 = 0.99$
       Extremely high risk

III.7. $Y = -1.5 \times 120 + 0.93 \times 0 + 3.3 \times 40 + 64.3 \times 0.2 = -46.7$
       High risk because $Y < 1$.

III.8.   Stand resistance calculation:
       PGR = I5/(I10 − I5) = 0.25/(0.4 − 0.25) = 1.67 (see Fig. 7.2)
       SHR = crown competition factor/% lodgepole pine basal area = 92/100 = 0.92
       Resistance = PGR/SHR = 1.67/0.92 = 1.82
       Phloem thickness variable = 60
       This stand falls in the low-risk zone because it is highly resistant (Fig. 7.4)

III.9. $\log_e Y = -0.681 + 0.447 \times 1 - 0.021 \times 1 + 0.505 \times \log_e (100)$
             $+ 0.487 \times \log_e (100/40) + 0.274 \times \log_e (100)$
          $= -0.681 + 0.447 - 0.021 + 2.33 + 0.92 + 1.26 = 4.39$
          $Y = $ antilog$_e 4.39 = 80.5\%$ defoliation

III.10. Estimated timber value lost   = (Mortality + growth loss) × green stumpage value*
                                          = (656 + 240) × 56 × 1000
                                          = 50,176,000
       *Green stumpage value/MMBF = green stumpage value/M × 1000
       Estimated values recovered   = Expected salvage volume × infested stumpage value
                                            = 470 × 40 × 1000
                                          = 18,000,000
       Timber benefits of control   = 50,176,000 − 18,000,000 = 32,176,000

Resource	Benefit of control
Timber	32,176,000
Reforestation	2,630,000
Fire protection	13,130,000
Wildlife habitat	0
Recreation	1,460,000
Range useage	−101,000
Total Benefit	49,295,000

Cost of control = cost/acre × number of acres infested

$$= 4 \times 331,000$$

$$= 1,324,000$$

Net benefit = 49,295,000 − 1,324,000 = 47,971,000

Harvest date of the timber was not considered; i.e., much of the growth impact may have been recovered by harvest because of increased growth in surviving trees. Wildlife habitat probably increased in a similar manner to range because of increased forage production beneath the defoliated overstory, providing a net benefit in terms of big game yields (in fact this area contains one of the largest elk herds in the country and considerable benefits to the local economy may have resulted from increased hunter success). Also, benefits from wildlife and cattle should be accumulated over the total rotation period.

Recreation impact is probably overestimated. Most studies indicate that recreationists have particular objectives, such as hunting and fishing, which are not deterred by defoliator damage.

PART IV: PRACTICAL EXERCISES

IV.2. *Douglas-fir Tussock Moth*

Structural diagram should include direct (−) density-dependent and delayed (−) density-dependent interactions with the food supply, and (−) delayed feedback with parasitoids and virus.

Death rates should remain low until populations get very dense and should then rise rapidly as starvation and virus disease become prevalent. Birth rates should also decline at high population density.

IV.3. *European Woodwasp*

Structure of the insect–plant interaction network should be similar to Fig. 11.3, with two negative and one positive feedback loops. You should also have delayed density-dependent (−) feedback loops linking the *Sirex* population to is nematode and ichneumonid parasite populations.

Sirex death rates will first increase with population density as larvae become crowded in the few weakened trees available. Death rates should then fall as larger populations cause stress to increasing numbers of healthy trees. At some point, however, ichneumonid parasitoids will begin to respond to increases in their woodwasp food and the death rate of *Sirex* will begin to rise again. At very high population densities the death rate will rise even faster as the tree population is depleted.

Sirex birth rates should be relatively constant at first but then begin to decline when the nematodes start to respond numerically to the density of their host. The birth rate should decline even more rapidly at very high densities because of the effects of crowding on female fecundity.

Note: The result of this deductive analysis is very sensitive to the siricid densities that induce numerical responses in the parasites. You should examine different response densities.

Diagnosis should involve an analysis of stand vigor (Chapter 7) and the incidence of parasitism in the resident population.

Prescriptions may involve silvicultural treatments or augmenting parasites, particularly the nematodes, which can be mass produced relatively cheaply.

AUTHOR INDEX

SUBJECT INDEX

TAXONOMIC INDEX